Copyright © 2025 Charlotte Chun | Dr Chun Coaching

All rights reserved.

No portion of this publication may be reproduced, distributed, or transmitted in any form or by any means without written permission from the publisher or author, except as permitted by US copyright law. For permissions, contact the author at: contact@earliestmemoriesbook.com

ISBN: 979-8-9991234-0-4 (Paperback)

Cover images designed by Charlotte Chun, M.S., Ph.D.

The book cover and chapter illustrations were designed by the author on Canva Pro by adapting elements from the following artists, in alphabetical order: *Amelia from Rimela Studio, Alyssa Babasa, Artist4love, Bakuden Creative, Clker-Free-Vector-Images, Color Vectors, Dan Putra, Dyaharum Pungki Revitasari, EasyToPrintArt, Geneo Yusuf, Gishella Rose, Giuseppe Ramos, Goodstudio, Gülhan Özcan, Harshita, Iryna Harmash, Jarirenjana, Jenny Lipets, JoyImage, Liliia Polos, Memed Nrh, Miadigital, Mickiiz, Milatoo, Murphyenormity, Natee Jindakum, Nays Design, Nlchoirina, Notionpic, Ouch! Illustrations, Pavelnaumov, Pixabay, Pri Patricio, Procrea, Rita Juwita, Sceptical Cactus, Sketchify, Stella Butalid, Sumil Shara, TiTi Lee, Ummu Zakir, Vectortradition, Vesvocrea, Zabi Jose, Zuperia.*

First Printed Edition: 2025

Published by Dr Chun Coaching

502 W. 7th Street

Suite 100

Erie, PA 16502

www.earliestmemoriesbook.com

EARLIEST MEMORIES

Discover the Psychology of Memory,
Explore Cross-Cultural Stories, and
Reconnect with Your Past

Charlotte Chun, M.S., Ph.D.

Contents

Chapter 1: Introduction 1
    Overview 1
    What to Expect from This Book 3
    What Is Your Earliest Memory? 4

Chapter 2: The Science of Memory 5
    Overview 5
    Foundations of Memory Processing 5
    The Grand Egg Fiasco 6
    Memory Reconstruction: Archaeologists of Our Personal History 10
    Forgetting: Common Memory Errors 11
    When Forgetting Is Helpful 13
    False Memories 13
    Extraordinary Memory Abilities 15
    Psychological Factors That Influence Our Memory 17
    Closing Thoughts 18

Chapter 3: Psychological Perspectives on Memory Development .. 19
    Overview 19
    Foundations of Memory Development 19
    Memory and Identity 23
    Collective Memory 25
    Closing Thoughts 29

Chapter 4: A Collection of Earliest Memories 30
    Overview 30
    A Collection of Earliest Memories from Around the World 30

Chapter 5: Emotional Memories 74
    Overview 74
    Unpleasant Emotions in the *Earliest Memories* Collection 75
    Pleasant Emotions in the *Earliest Memories* Collection 78
    The Role of Emotions 80
    Emotions and Social Bonds 81
    Emotions and Memory Processing 82
    Emotional Quality and Intensity 84
    The Impact of Stress on Memory Processing 86
    Somatic Memory 89
    Closing Thoughts 90

Chapter 6: Sensory Memories 92
    Overview 92
    Sensory Memories 92
    Sight 93
    Sound 97
    Smell 99
    Taste 103
    Touch 104
    Closing Thoughts 107

Chapter 7: The Significance of Our *Earliest Memories* 108
    Overview 108
    Earliest Memories Collection Themes 108
    What Makes Our First Memories So Special? 115
    The Nature of Nostalgia 118
    Closing Thoughts 119

Join the *Earliest Memories* Community 122
    Share Your Earliest Memory 122
    Connect with Us on Social Media 122
    Get the Word Out 122

References 123

About the Author..143

Acknowledgments

My most sincere gratitude goes out to everyone who shared their earliest childhood memories with me over the years. From friends and family to paid respondents to strangers who came across my business card, this book would not have been possible without your contributions. Thank you for sharing these precious, intimate moments in your life!

My heart is filled with appreciation for my parents: to Phil Gates for supporting my "nomadic writer's retreat," and to Maureen and Tom Welch for graciously hosting me during the final stages of writing. Thank you to Matt Gatner for your insights on the parent-child interaction, and to Heidi Kern Hannan and Maureen Welch for your valuable feedback on the book cover.

Finally, I want to thank my top-notch editors, Sabine Huber, who provided scientific content editing, and Thomas Symalla, who provided copyediting. Your knowledge and talent helped transform the book into something I am delighted to share with the world!

For Gram, the one we could never forget.

Margaret "Helen" Luley (1923-2022)

"One is always at home in one's past..."

-Vladimir Nabokov

*Speak, Memory: An Autobiography Revisited*₁,

1966 (p. 116)

1

Introduction

Overview

Our memory takes so many different forms—we can drum up obscure facts for a game of trivia, remember how to do a cartwheel after years of hiatus, or hold a phone number in mind while we search for a pen. We may even surprise ourselves by still knowing all the words to a popular '80s hit! But there is something very special about the memories we have from our own life experiences.

Many of us take the profound gift of our memory for granted, perhaps until someone dear to us has trouble remembering things or our own memory starts to falter. We all rely on our memory constantly throughout the day, but we don't always pay attention to how often we use it. You might suddenly become aware of your memory's limitations when you struggle to recall a friend's birthday or that pesky password for a rarely used website. However, most of us tend to gloss over the massive amounts of data that are stored in our brain. For example, we don't pay attention to the fact that we remember our pets' names every morning or that we know where the canned tomatoes are in the pantry.

Now, some people are clearly better at remembering things

than others—are you the person who remembers exactly what outfit you wore to brunch six weeks ago or are you the constant forager who never knows where you left your keys? To some extent, these differences can be explained by context. For example, it makes a big impact whether you have systems in place to help you remember things… and whether you actually use those systems (like putting your keys on the key rack). But there are also individual differences in memory that are quite remarkable, such as how many things you can remember at once, how easily you can recall someone's name, how much detail you can drum up about a past event, how long it takes you to forget something, and how far back your memories reach.

What to Expect from This Book

Memory is a beautifully complex phenomenon, and there are more than enough books on the topic to fill an entire library! This book will focus on our earliest memories and what makes those first traces of memory so special. We will examine memory from a scientific and philosophical perspective, written by a psychologist with a background in neuroscience and cognition. Don't worry if you don't know much about the brain (you're in good company… Scientists don't, either!). I'll share research and other ideas on memory using language that's easy to understand.

In Chapters 2 and 3, you will gain a basic understanding of how human memory works and how it changes over time. In Chapter 4, you will read a treasured collection of people's earliest memories from across the world. In Chapters 5 and 6, we will discuss the emotional and sensory nature of these memories. Finally, in Chapter 7, we will explore the themes that arise from the *Earliest Memories* collection and why certain memories stay close to our hearts.

This book will not cover memory in relation to neurocognitive disorders, post-traumatic stress disorder (PTSD), neurodivergence, or other clinical cases. In those cases, memories are processed in different ways than usual, so those topics warrant their own deep dive. This book focuses primarily on topics related to non-traumatic childhood memories, although there are some stories of illness, injury, and natural disaster in the *Earliest Memories* collection in Chapter 4. There is also discussion of a forensic case related to assault (without going into details of the incident) in Chapter 5 that may be upsetting for some to read.

What Is Your Earliest Memory?

Before you read any further, take a moment to think of your earliest memory from childhood. Don't worry if you're not completely sure—just close your eyes and see what comes to you instinctively.

What happened in that moment and what did you experience with your five senses? Where did this experience take place? About how old do you think you were? How confident are you in that timing? What gives you the sense that this is the earliest memory you have? Why do you think this particular memory has stuck with you after all these years? And what meaning do you draw from it at this point in your life?

If you want to share your earliest memory to be included in our collection, you may email it to: contact@earliestmemoriesbook.com. I encourage you to keep your earliest memory in mind as you read the upcoming chapters and see what rings true with your own experience. Will your memory fall into one of the common themes covered or is it in a category all its own? Are there any similarities between your earliest memory and one from another country? Do you think you were older or younger than the average age of people's first memories? Read on to find out!

2

The Science of Memory

Overview

This chapter covers fundamental concepts about human memory and its influence on our daily lives. We begin by describing the basic science of how our memory works. We discuss the dynamic nature of our memories and how they are constructed and reconstructed over time. We explore the spectrum of memory functioning, from everyday forgetting to extraordinary memory abilities. Finally, we examine self-concept and how our subjective reality shapes our perception of the past.

Foundations of Memory Processing

With a topic as complex as memory, I'm sure you won't be surprised to hear that there are many ideas out there to explain *memory processing*. Memory processing is how our memory works, or the way that our brains take in, store, and retrieve the information we remember. I'll break down some widely accepted scientific theories on memory processing into simplified terms.

This book focuses specifically on *autobiographical long-term memory*, or put simply, the memory of our own experiences. Autobiographical memory includes our memory of personal events and other information that relates to us.[2,3] Long-term memory is the part of our memory system that retains everything we remember for more than about 20-30 seconds.[4] So, whether it's something you remember from just a minute ago or your earliest childhood memory, both are stored in your long-term memory.

Before we dive into the details of how our long-term memory functions, let's consider a real-life example to illustrate.

The Grand Egg Fiasco

Imagine you get home from a last-minute grocery run for a dinner party you're hosting. As you are walking up the stairs, you reach for your keys, and one of the bags you're carrying tilts to the side. The carton of eggs on top tumbles to the ground and your hopes of making quiche for dinner are crushed on the concrete steps!

As if the Humpty Dumpty reenactment wasn't jarring enough, you also must explain to your dinner guests later that night why their favorite mushroom quiche is no longer on the menu… and then try to convince them that mushroom soup is really better anyway.

There is a lot happening in your brain during this experience! Let's break it down in terms of memory processing. The way we form memories and recall them later happens in three main stages: *encoding*, *storage*, and *retrieval*. Encoding is the process of translating the information we take in through our senses into a code that our brain can work with. Storage is the process of preserving that coded information within our memory system. Retrieval is the process of accessing that stored information and activating it for use.[3-5] Here we'll explore each stage of the process.

① *Encoding*

During the first stage of memory processing called *encoding*, the details of what we experience are received and translated by the brain into a form it can process and store. In the example of the "The Grand Egg Fiasco," your brain captures the context of when and where the unfortunate experience occurred, what happened, and in what order the events unfolded. The important details are translated for storage, such as the sound of the crunch, the sight of the yolk oozing onto the concrete, the meaning you place on the eggs breaking, and your emotional response.

The internal context (including your mental and emotional state) and external context (what environment you are in and who you are with) are encoded as part of your memory of what happened.[6] All these details become associated with one another, blending together as one holistic experience.[7] Our brain integrates the sensory information, our interpretation of this information, and our emotional experience into a pattern of meaningful connections.

Personal experiences are not encoded like cut-and-dry facts. Rather, we actively construct our memories. What we remember later is colored by our thoughts and feelings about what happened and our broader belief system. Ultimately, it is our own unique understanding of the event that will be stored as a memory in the next stage of processing.[3,8]

 Storage

During the second stage of memory processing called *storage*, coded information is saved and integrated into our long-term memory. Scientists think that memories are stored in pattern-like connections between different brain cells. When a new memory is formed, it creates new connections between cells in key parts of the brain. These connections help store and organize the information so that we can retrieve it later when we need.[9-12]

Stored memories are incorporated into complex frameworks of our existing knowledge. For example, in "The Grand Egg Fiasco," the memory of the eggs breaking will be integrated with your understanding of the laws of gravity, the repercussions of rushing, and your self-perception around how you navigate challenges. Unless an event radically shifts our perspective, we tend to store memories in a way that is consistent with our existing beliefs.

3 *Retrieval*

During the final stage of memory processing called *retrieval*, our brain finds and accesses the stored memory. Returning again to "The Grand Egg Fiasco," when you tell the story to your party guests later that evening, you will need to retrieve the memory from storage. As mentioned earlier, memory storage creates new patterns of connections among brain cells. When something cues us, or prompts a stored memory, those patterns of interconnected brain cells become activated. The communication among these cells allows the relevant details of the memory to be retrieved and integrated so that we experience one cohesive memory within a moment's time.[4,5,13,14]

Let's use a visual example. When you look at a familiar face, cells within parts of your brain that are specialized in vision become activated by the visual cues. Then these cells respond by activating, or communicating with, other parts of the brain that are needed to recognize the person's face and retrieve your memories of them. For instance, if I show you a photo of Michael Jackson, it will activate different patterns of brain cells that have previously stored memories associated with him. This evokes your concept of Michael Jackson, which may include both positive and negative associations.

The more you are exposed to an experience repeatedly, the stronger those associations among patterns of brain cells become (*"Donuts are delicious, donuts are delicious, donuts are delicious."*). Remembering something over and over again is like driving your car down the same dirt path every day. Eventually, the tire tracks wear a rut in the road and you begin to retrace the same path automatically.

Context also plays a pivotal role in memory retrieval. Contextual factors are initially captured during memory encoding. We recall things better when the context during retrieval matches the original context from encoding. This phenomenon was famously demonstrated by the scientists Godden and Baddeley in the 1970s[15] when they conducted a creative experiment to test their theory. They asked deep sea divers to memorize a list of words, with half of the group on the beach and the other half underwater. Then they tested their memories, some in the same context where they learned the words, and some in the opposite context. The divers who memorized words underwater recalled more words when tested underwater versus on the beach, and vice versa. The context (land vs. sea, in this case), serves as a cue that impacts what we remember.

Context is not the only factor that impacts memory retrieval. Like encoding and storage, retrieval is not a simple, objective process. There are many psychological influences at play when we retrieve a stored memory, as well as the potential for errors and biases, which we'll cover in the next sections.

Memory Reconstruction: Archaeologists of Our Personal History

When we remember our life experiences, we engage in a type of "mental time travel" to restore the information we have from that timeframe.[16-18] However, revisiting a memory is not like replaying a recording of the event. We can't remember events exactly as they occurred, so we must rely on other clues to recreate the memory.[19,20] For example, if we are retrieving the memory of a past doctor's visit, but don't remember all the details, our brain tends to fill in the gaps with our knowledge of what typically happens at a doctor's office. We reconstruct our memories by piecing together information from that context that aligns with our worldview.

Just as we actively construct the memories that are formed during encoding and storage, we reconstruct our memories each time we retrieve them. *Memory reconstruction* is the process of piecing together the different traces of what we experienced to yield a memory that feels cohesive with our understanding of ourselves, others, and the world around us.[13,14] As cognitive psychologist Ulric Neisser famously put it, "Out of a few stored bone chips, we remember a dinosaur" (Neisser, 1967, p. 285).[13]

Our memories are not etched in stone—they are quite dynamic! They settle into our long-term memory over time, especially during sleep.[21] Memories can be updated each time we think about them. They can shapeshift over time based on how we see the world and, particularly, how we see ourselves.[3]

Years later, who knows if you will recall "The Grand Egg Fiasco" leading up to your dinner party? But if you do, chances are, you may see it in a different light. Have you ever looked back on an experience with "fresh eyes?" Over time, feelings of disappointment or embarrassment can lose their sting as we gain perspective and make meaning of events. When we revisit an old memory with new wisdom, our mind updates that memory to become more aligned with our current understanding of life. Our personal growth permeates not only our present and our future experiences but also the way we view our past.

Forgetting: Common Memory Errors

As you can see, remembering is quite a complicated process!

Errors can happen at any point during encoding, storage, or retrieval. Thus, there are many reasons why we might forget something, or fail to remember it correctly. Let's consider several examples.

First, information may be encoded inaccurately from the moment we experience it. We can only encode what we pay attention to, so if we are distracted during a meeting, for example, we may not remember what the presenter said because the information was never fully received and encoded in the first place.[4,22]

Memories can also be stored improperly. For instance, if we are under stress or are sleep-deprived, newly encoded information might not be integrated properly into our existing knowledge structures. Without proper storage within those patterns of associated brain cells, the unstable memory can easily be disrupted or lost.[4,23]

Finally, we have retrieval errors. How do we account for memories that were once stored but cannot be retrieved? There are multiple theories on the subject. Some scientists believe that memories can simply fade over time as the connections among brain cells weaken with infrequent use. Others propose that we never fully lose information once it is stored in our long-term memory—we simply lose access to it or are unable retrieve it without external reminders to prompt our memory.[3,22]

Additionally, the intake of new information can interfere with the retrieval of a similar memory.[22] For example, it might be difficult to remember the band name *The Smiths* if you recently heard a song by *The Shins*. Both band names are stored in your long-term memory, but when you hear a song by *The Shins*, it activates those associations. This exposure makes your memories of *The Shins* stronger and more accessible. Consequently, when you try to recall a band name, *The Shins* are top of mind. Since the name sounds similar, it can compete for your memory, interfering with retrieval of *The Smiths*.

When Forgetting Is Helpful

There is no way around it—our memory is not perfect. We've all experienced the impact of faulty memory in our day-to-day lives. You think you can remember a short grocery list, but the cilantro slips your mind. You thought you texted someone back but, it turns out you never actually replied. Your friend's brother greets you at a barbecue, but you completely forget his name. These experiences are a dime a dozen!

As frustrating as these memory lapses can feel, forgetting can also be adaptive. Forgetting helps ensure that old information doesn't get in the way of new things we need to remember. For example, if you clearly remembered every spot where you parked your car in the past, it could cause major confusion about your current parking spot. Scientists now understand that information does not just passively disappear from our memory over time; we also engage in "active forgetting" as our brain blocks irrelevant memories from coming to mind. This helps us stay focused on the task at hand so we can be more efficient in our daily lives.[5,24] So, not all forgetting is a bad thing!

False Memories

In addition to forgetting, we are also prone to conjuring up false memories. In scientific experiments, participants will commonly say they recall events that never actually occurred, especially under certain circumstances. False memories are more likely when the context seems to support that memory, when someone suggests it, and when the event seems believable.[25-27] For example, when prompted, people often claim to have seen videos of high-profile news events, even when no real video recording exists. In one study, nearly half of the participants said they had seen film of Princess Diana's car crash, although there is no footage of that tragic event.[28]

False memories occur in our day-to-day lives, perhaps more often than we might think. Overall, our memory is generally reliable, but it is not infallible. We usually do not fabricate memories out of thin air; however, we may not recognize when our memories are several degrees from "the truth."

Let's take the all-too-common example of a miscommunication with your partner about upcoming social plans. You *swear* you told them about Sunday brunch (you remember the conversation!), but they insist you never mentioned it. We're often quick to assume that we are right and the other person is mistaken. But in reality, our minds can form false memories that we're convinced are real.

False memories can result from influences at any stage of memory processing: encoding, storage, or retrieval. For example, you may have discussed weekend plans with your partner intending to tell them about brunch, but your friend was texting you at the same time. With your attention divided, you never actually mentioned the brunch to your partner. The distraction from your phone disrupted the encoding process and your brain did not capture a complete representation of the conversation.

Later that day, you were chatting with friends about the upcoming brunch and the two conversations were conflated during storage.

During the retrieval of the conversation with your partner, your mind fills in the gaps of what was said. You intended to tell them about brunch, so this expectation dominates your perception, and details that were not part of the original conversation are incorporated into your memory. Since you know you mentioned the brunch *(just not with the person you thought...)*, you might feel confident in this false memory. This is how misunderstandings can happen and how two people with good intentions can spend Sunday morning arguing instead of sipping sangria in the sunshine.

Extraordinary Memory Abilities

Now let's turn to the other end of the memory spectrum. Imagine being able to recall your life experiences with remarkable precision... If you could take a pill and remember everything you experienced throughout your life, would you do it? Some people are born with a rare ability called *hyperthymesia*, or highly superior autobiographical memory (HSAM).[29] Individuals with HSAM have an extraordinary memory capacity. They can provide dates and specific details of their own lives and public events with extreme accuracy. For example, they can tell you who they ate lunch with on a random afternoon 12 years ago, what they wore, where they dined, and what food they ordered. They remember this information automatically, without the use of any special memory techniques.[29]

Personal accounts from individuals with HSAM are quite fascinating. They generally view their strong memory as beneficial. However, for some, it can be a double-edged sword. A superior memory can sometimes trigger interpersonal conflict. Think back to the brunch example. What if your partner was always the one insisting they remember correctly? Of course, some people with HSAM are quite satisfied with their relationships. But anecdotally, many report struggling to maintain long-term partnerships. It can be difficult for people with HSAM to share their extensive memories in a way that others can relate to. For some, it feels alienating not to know anyone else with this ability. Take this quote from an interview on CBS News's *60 Minutes* (Finkelstein, 2011[30]):

> Sometimes, having this sort of extreme memory can be a very isolating sort of thing. There are times when I feel like I'm fluent in a language that nobody else speaks. Or that I'm walking around and everybody else has amnesia.

Further, for some individuals with HSAM, the constant presence of the past makes it difficult to fully embrace the present moment. For example, consider the following excerpt from a segment on NPR's *All Things Considered* (Spiegel, 2013[31]):

> In her life, there are no fresh days, no clean slates without association. Every morning when she wakes up, details of that date from years before are scrolling through her mind, details that can profoundly affect the new day she's in. [...] Because the past is so viscerally right there, so available, she finds that when the present gets overwhelming, it's hard not to retreat to the past.

For all the ways that HSAM differs from normal memory, there are also similarities. Despite their extraordinary autobiographical memory, research has shown that people with HSAM perform about the same as the general population on common memory tests, like remembering lists of numbers and words. Their superior memory is primarily related to calendar dates and personal events.

It is also important to note that they *do not* have "perfect memory." People with HSAM retain many more memories and with far greater accuracy than the rest of us, however, when recalling events, they still rely on the typical memory processes to reconstruct a narrative of what happened.[26,29] As we explored earlier, reconstructed memories are based on our own subjective understanding. HSAM or not, memory reconstruction makes us all susceptible to inaccurate and distorted memories. It is simply part of being human.

Psychological Factors That Influence Our Memory

There are underlying psychological reasons why we remember things in a certain way. For one, we are privy to our own internal world, so we often project our intentions onto the outcome. For instance, if my goal was to be helpful, I might remember working very diligently on a group project, unknowingly exaggerating the memory of my contribution. We are sometimes unaware when we edit or embellish our stories—we really believe that's what happened![32]

Another reason why we remember things a certain way is to maintain a positive self-concept. For most of us, it's important to our emotional well-being to view ourselves in a positive light,[33] although this does vary by culture. For instance, in Japanese culture, the need for positive self-regard may not be as strong as in other cultures.[34] In Western society, people tend to see themselves and their past actions through rose-colored glasses.[34,35] For example, we may selectively remember times when we performed well at work, ignoring the times we made mistakes. As we'll discuss in the next chapter, the way we enhance ourselves in our memory plays a key role in shaping our identity over time.

Closing Thoughts

In this chapter, we explored many examples that highlight the subjective nature of long-term memory. You may be wondering what this means for your own memory and your life. My goal is not to make you discount or distrust your memory—it is undeniably one of your most useful tools! Rather, I believe it's important to recognize your memory for what it is: a representation of who you are and what you believe overlaid onto your experience of the past. As you will see in the next chapter, this is a two-way process: our self-concept colors our memories of the past while our memories of ourselves form a foundational piece of our self-concept over time.

3

Psychological Perspectives on Memory Development

Overview

Now that you have a basic understanding of how memory works, we will examine how our memory changes from early childhood to adulthood. This chapter explores how memory is intertwined with our cognitive, social, emotional, and identity development.

Foundations of Memory Development

It is quite uncommon for people to remember things from before the age of three, though there are individual differences in what we recall from childhood.[20,36] One researcher conducted a meta-analysis of over 11,000 cases of adults recalling their "first memories" and found that there were almost no memories recounted before age three. In that study, the average age of participants' first memory was 3.4 years old.[36] Memories within the three- to five-year age range are often described as "fragments." In general, detailed long-term memories can be stored more reliably starting around age five.[22,37]

Beginning around age three, children can remember personal experiences temporarily, but most of those memories will be forgotten over time. This phenomenon is often referred to as *childhood amnesia*.[38] It is difficult for young children to form stable long-term memories because they do not have a way to organize events into a mental model or representation of their world. They have not yet developed the necessary cognitive frameworks nor the complex connections among brain cells that are needed to support long-term memory storage.[38-41] Children need exposure to new ideas, environments, and life experiences that will allow them to create connections among brain cells, forming networks of associations. You can think of those networks as an ever-expanding web that new memories can weave into by integrating with the child's existing knowledge and experiences.

The acquisition of language is also instrumental in strengthening memory development in young children. Without language, young children may struggle to encode, store, and retrieve experiences effectively in their memory. Language allows us to label, understand, and form narratives about events in a way that makes it easier to process, mentally organize, and recall the information.[22,42]

Imagine, for example, a young toddler who feels frustrated because they can't screw the top back on their sippy cup. They might express their emotions by stamping their feet or crying but they are not yet able to describe this feeling with the word "frustration." With limited vocabulary to describe their experience, it may be challenging for the toddler to remember what happened later.

Putting words to our feelings enables us to form stronger memories by helping to make sense of our experiences and form organized connections in the brain. For children's memory to fully develop, they need language and a mental model of how the world works. Without these supporting structures in place, it is difficult to form lasting memories.[22] It would be like hanging a large ornament on a tiny Christmas tree—the supporting framework simply is not strong enough!

As children age and develop, their encoding, storage, and retrieval abilities all improve,[43,44] with the greatest changes thought to occur in the ability to store information in long-term memory.[38,44] As children's memory abilities continue to evolve, parents play a crucial role in bolstering this development. They can provide a supportive framework, often referred to as a "scaffold," to help their child recall things. Strategies such as asking questions, guiding the conversation, and encouraging further elaboration about experiences makes the task of remembering easier for children.[41,45,46] With additional support from a caregiver, children can recall things that would typically be too complex for them to remember. This allows them to practice and engage with memories that would have been otherwise inaccessible. Here's an example of a parent supporting their child's memory:

> **Parent:** "Are you excited to go to the park?"
>
> **Child:** "Yeah!"
>
> **Parent:** "Remember last time we went on the big-kid slide?"
>
> **Child:** "Yeah."
>
> **Parent:** "How did it feel?"
>
> **Child:** "It was scary."
>
> **Parent:** "It felt scary, huh? But you were so brave, and you slid down anyways."
>
> **Child:** "Yeah, I went fast."

Parent: "You did! And what else did we do at the park?"

Child: "We got ice cream!"

Parent: "We got ice cream, that's right. And do you remember who got ice cream with us?"

Child: (silence)

Parent: "Was it… Lisa?"

Child: "No."

Parent: "Was it Grandpa?"

Child: "NO!"

Parent: "Aunt Erin?"

Child: "AUNT ERIN!"

Parent: "That's right, it was Aunt Erin!"

When parents engage with a child to reminisce and elaborate on memories, they can positively impact the accuracy and capacity of their child's memory. These interactions also promote children's broader cognitive, social, and emotional development. Reminiscing strengthens the connections among brain cells that are associated with the memory and helps children mentally organize events.[47,48] These conversations also cultivate children's language expression by encouraging them to practice and expand their language abilities and narrative structures.[42,48,49]

Further, reminiscing about emotional experiences can help develop children's social and emotional skills. Talking about emotional memories can strengthen social bonds, improve conversational skills, encourage open discussion about pleasant and unpleasant emotions, and teach children to understand and cope with difficult experiences.[48,49] The ability to remember and understand one's own experiences helps lay an important foundation for children's development of self-identity and aspects of their social and emotional competence.

Memory and Identity

Psychologists and philosophers alike have proposed theories to account for the link between memory and self-identity. One theory suggests that the ability to remember depends on having a personal identity: To have a mental framework of things happening to me, I must first have a concept of "me." Once that self-concept is present, memory can connect past experiences to current awareness.[50] Others believe that a "sense of self" forms over time as we create a story from what we remember about ourselves. We pull a "narrative thread" that connects us to our past, present, and future.[51] These ideas are brought to life beautifully by renowned novelist Vladimir Nabokov in a passage from his memoir, *Speak Memory* (Nabokov, 1966, p. 21[1]):

> I see the awakening of consciousness as a series of spaced flashes, with the intervals between them gradually diminishing until bright blocks of perception are formed, affording memory a slippery hold. I had learned numbers and speech more or less simultaneously at a very early date, but the inner knowledge that I was I and that my parents were my parents seems to have been established only later, when it was directly associated with my discovering their age in relation to mine.

Nabokov describes a memory of his mother's birthday when he asked his parents how old they were. Upon learning their ages, he suddenly grasped the idea that they were alive before he was born, which triggered a shocking realization (Nabokov, 1966, pp. 21-22[1]):

> At that instant, I became acutely aware that the twenty-seven-year-old being, in soft white and pink, holding my left hand, was my mother, and that the thirty-three-year-old being, in hard white and gold, holding my right hand, was my father. Between them, as they evenly progressed, I strutted, and trotted, and strutted again, from sun fleck to sun fleck, along the middle of a path [...]. Indeed, from my present ridge of remote, isolated, almost uninhabited time, I see my diminutive self as celebrating, on that August day in 1903, the birth of sentient life.

Nabokov's realization of his place in time and sudden perception of himself as a distinct being established a sense of self that would tie together the rest of his life's experiences. These passages illustrate how interdependent memory and consciousness are. As we discussed in Chapter 2, our self-concept influences our memory, acting as a subjective lens through which we view our past. At the same time, these memories become part of the dynamic portrayal of our self-concept as our memory and identity continually shape each other over time.

Collective Memory

Our individual memory is just one piece of the puzzle. Social influences on memory typically begin with our caregivers and family and continue to evolve over time as our social network expands. *Collective memory* includes the shared recollections of a group that extend beyond any one individual.[52] As we embed ourselves in different social groups, we can tap into and even shape the collective memory of the communities we become a part of.

I'll share an example from my own life to illustrate the role of individual vs. collective memory. For a couple of summers in college, I worked as a counselor at Camp Hardtner, a small yet magical place nestled in the pine trees of Louisiana in the Southern United States. Let me tell you, playing four-square in a humid pavilion with no air conditioning in the dog days of July was no joke! I can almost feel the sweat dripping down the back of my knees just thinking about it… but we always found creative ways to beat the heat on a budget. From playing Battleship water balloon games to ice cube relay races, I bonded with the campers and my fellow counselors and made many meaningful memories those summers.

After college, I moved away, but whenever I would get together with my old camp buddies, reminiscing about our zany adventures was always the preferred pastime. Most of us had also attended the camp as children and we were still telling stories of those epic experiences from many years past. There was something so precious about reliving that golden era, sacredly preserved in time away from the hustle and bustle of our adult responsibilities.

As years went on, however, I started to notice something unsettling. I always thought my memory was fairly strong, but over time, I would hear my friends recount details I could have never conjured up. Or even worse, they would tell stories they all seemed to know but I had almost no recollection of. I began to hear the dreaded phrase, *"I can't believe you don't remember that!"* What was happening? Why was I losing my cherished camp memories?

After moving away, I became disconnected from the physical and social context of those experiences. I only occasionally dusted off my camp memories during sporadic phone calls or the rare group reunion. As time went on and those memories were not activated regularly, the connections began to fade. Without access to the context or cues that would prompt retrieval of those memories, it was easy to forget specific details or even entire events. By the time I heard some of the camp stories again, the traces of those memories were too faint and could not be retrieved.

In contrast, my friends in Louisiana got together regularly and continued to tell tales of those camp summers. By repeating these stories as a group, they reinforced and expanded those memories. Every time a story was told, details that one person or another may have remembered were incorporated into the fold, contributing to a collective camp narrative. While one person's memory may naturally fade, collective memory can grow and transform, taking on a beautiful life of its own as stories are passed down over time and turn into folklore.

On a global level, collective storytelling, especially through oral traditions, is an important part of every culture.[53] *Cultural memory* represents the shared knowledge, history, traditions, values, and narratives that are passed down within a culture. Whereas collective memory belongs to a specific group or community, cultural memory belongs to a broader culture or society. Cultural memory serves as a vital mechanism for preserving information, including collective trauma and triumph. Beyond that, it also fosters community connection, creating a sense of belonging and shared purpose that extends beyond individual identity.[54,55] We long to feel part of something bigger than ourselves.

National Geographic calls cultural memory "the constructed understanding of the past."[56] Once again, we are reminded that memory is not a robotic recounting of facts, but rather a dynamic retelling of the past, whether by an individual, group, or society. Our memories are continually reconstructed according to the meaning they have for our lives in the present moment and as we look to the future.

Closing Thoughts

In this chapter, we discussed how memory changes during childhood and some of the key factors that impact children's memory, such as brain development, language acquisition, and caregiver support. We further examined the convergence of individual, social, and cultural dynamics that impact our memory and shape our identity. In the next chapter, you will read a collection of earliest childhood memories from 20 countries across the world. As you enjoy a peek into these intimate moments from people's childhoods—precious and unpleasant, profound and unremarkable—notice the themes that arise. What common experiences and universal emotions can you find that capture the essence of what it's like to be a child and what it means to be human?

4

A Collection of Earliest Memories

Overview

In this chapter, you will read 100 earliest memories collected over many years from family, friends, strangers, and paid respondents from all around the world. These memories were curated to bring the topic of childhood memory to life, to celebrate the diversity of memories from various cultures, and to create a global platform where people can share their first memory.

The memories have been lightly edited for spelling, grammar, and clarity. They have been anonymized and are only identified by age and location, which reflect each person's best guess of when and where their first memory occurred. Please enjoy these special moments in the lives of those who were generous enough to share a peek into their childhood.

A Collection of Earliest Memories from Around the World

My earliest memory is reaching through the bars on my crib toward my twin sister, who was in the next crib. I think the cribs were painted white with spindle bars, and the room was sunlit, like afternoon, but it could have been morning. My sister looked up and reached back and we mushed hands as a greeting or

something. As a grown woman, I would now characterize that as our first secret handshake.

Age 2, Chalfont, Pennsylvania, USA

I remember waking up at home in our apartment in a playpen, feeling warm and sleepy. It must have been late afternoon because I could see a sharp line of sunlight across the opposite wall of the room and part of it was shining on me. I threw off my blankets and stood up at the edge of the playpen and looked around. I was alone, but not afraid. My toys were not on the floor and I found that puzzling. Where did they go while I was napping? I could hear the soft sound of my mother's voice somewhere else in the apartment. I think she was on the phone. I yawned and stretched. I felt content and a sense of waiting.

-Age 3, Rockville, Maryland, USA

My earliest memory is of my older sister bringing me chamomile tea while I lay in my crib in my room. My sister is ten years older than I am. The chamomile tea was in a bottle with a nipple. I really enjoyed the tea, and I tried to find ways to ask for more. The tea was sweet and warm. My sister always smiled and talked to me while she stood over the crib. I didn't understand what she was saying, but I enjoyed the positive energy and playfulness. I remember the texture of the blanket, the warmth and coziness, my stuffed teddy bear, and how I would touch his eyes and nose with my lips. I remember the bars on the crib, their shape, and touching them with my hands. Since the shelf was right next to the crib, I remember feeling the coldness when I touched it with my hand.

-Age 1, Belgrade, Serbia

I remember waking up in my bedroom and feeling odd and new in a way. It was almost like I had just woken up from a very, very long sleep. Everything around me felt familiar—my chest drawer in the corner of my room, my toys stacked together by my bed, and the transparent curtains that showed a glimpse of sunlight. I remember feeling everything. I stepped out of my bed and felt the cool, hard touch of the wooden floors. I walked out slowly and curiously into the hallway, brushing my fingers on the roughly textured wall, and straight to the living room where I could hear the faint sound of the television. I was aware of where I was going—there was no question at all. Then I saw my father. This is my first memory of him. He was sitting on the couch with his legs propped up, watching the morning news. I remember feeling so happy upon seeing him. He greeted me with a good morning. In turn, I flashed a big smile, ran up to him, and sat right on his lap as he continued to watch the news. I believe that this particular memory stuck with me because of how I felt at the moment—pure happiness. The feeling I had that particular day is different from any other. Even thinking about it now, it feels so pure and authentic, and I just want to go back to that moment.

-Age 7, La Trinidad, Benguet, Philippines

We were at home in the little house that we wouldn't occupy for much longer. My mom was sitting on a rocking chair in the corner of the living room. I was sitting on her lap. My father was standing nearby. My sister hadn't been born yet. I remember this because it was so comfy, so peaceful, and so happy. There was nothing else like it. It is the ideal that I have always sought to go back to and never could. To me, it is warmth, the picture of a complete family, and total innocence. It is the feeling of belonging. We may have been reading. It didn't matter because we were just together in that small space. It is a family memory of peacefulness and wholeness that couldn't last. It could never be recreated.

-Age 2, Eastlake, Ohio, USA

I remember my mom reading me a story one evening. I sat in her lap as she read. The story was about a boy whose mom had just had a baby and the changes that happened in the boy's life because of that event. After my mom finished reading, she talked to me about the new sibling I would soon have.

-Age 3, Paris, Texas, USA

I remember being in my room playing with some action figures and my dad picking me up. I remember pretending to be one of the X-Men before he came in. I remember his strength when he picked me up, and I remember giggling and laughing as he spun me around in circles. I used to think he was impossibly strong.

-Age 7, New Carlisle, Indiana, USA

My earliest memory is of waking up in the middle of the night with my eyelids crusted together so I couldn't see anything! I cried out in fear. Flash to my mom holding me in the bathroom, gently wiping my eyes clear with a damp washcloth and I can finally see.

In the bathroom mirror, I see myself snuggled up against my mom, and I am safe.

-Age 2, East Meadow, New York, USA

One of the earliest childhood memories I have is when my dad took me and my siblings fishing, and I caught a fish for the very first time. I think I remember this because it was a lot of fun and I felt a good connection with my father for the first time, and those imprints last forever.

-Age 4, Seal Beach, California, USA

In my childhood, I went to the lake with my father once. He let me handle the fishing rod and I caught a fish on that day, and it was the first one of my life. I was so happy and we enjoyed the day a lot. My mom cooked those fish for dinner.

-Age 12, Honolulu, Hawaii, USA

My earliest childhood memory is when I was around two or three and we were visiting my family down south. My grandpa had rented a house on the lake for us to go out on the boat, swim, and enjoy our time. During this time, I remember playing in the water, throwing toy cars around, and running around with my cousin who is the same age as me. I think that this memory stuck with me because it was the first time that we experienced being around a large body of water. This was also the first time I would go to my grandparents' house.

-Age 3, Florence, South Carolina, USA

I was in a sled being pulled down a snowy driveway by my grandpa. I was having fun imagining myself having more control than I did. I was two years old; given his date of death, it must have been 1978. The wind was cold on my face and I was bundled up as tight as the younger brother from *A Christmas Story* on his way to school. Gladstone, Michigan gets very, very cold in the winter. The sled my grandfather was pulling me in did not have a rope on it, but a long shovel handle made of wood that was attached to the front of it. The sled itself was orange plastic and my father was walking with us. I could not really understand what they were talking about, but their voices made me feel warm inside. My grandfather was breathing hard but had the biggest smile on his face.

-Age 2, Gladstone, Michigan, USA

I went to a cafe with my mum and my younger (by 20 months) brother. We sat on red leather seats and this was exciting for me, at age three, because it is the first time I had ever been to such a place. There was a glass partition between the seating area and the kitchen, and I vividly recall seeing two men working with a white stretchy material that I could not identify. I asked my mum and she told me that they were making bread, possibly for pizza. We did not have pizza—we had sandwiches, and my brother and I each had a glass of something fizzy. Cola? Orangeade? I felt very "sophisticated," even though this was not a word that I knew at the time. On the way home, we stopped at a grocery store and I felt proud that I'd just been to a cafe for the first time.

-Age 3, Worksop, Nottinghamshire, UK

My earliest childhood memory that stands out the strongest to me is from when I lived in New Orleans. I remember being small and my father put me on his shoulders to watch Mardi Gras. I remember the floats going by and people throwing bead necklaces from the floats. I remember my mother and aunt being dressed in costume. I can remember the excitement and all the sounds and bright colors. I think it stuck with me because it was such an exciting and dramatic day. It's still very vivid in my mind.

-Age 4, New Orleans, Louisiana, USA

My earliest childhood memory is my fifth birthday party, which happened at Chuck E. Cheese in Louisville, Kentucky, where I grew up. I remember having begged my parents to allow me to have my party there and to invite my class because that was one of the most popular places to be as a kid. I distinctly remember being in the main room where we ate pizza, and I got to sit with my friends who had come for the party. In that room, there were also musical animals that played songs. Afterward, we all had tokens that we used to play video games and carnival games. I remember also playing on the slides and in the ball pit. I was so excited to be there and have fun with my friends. We ran all over the place, playing games all day and periodically running back to our tables for more pizza. I was thrilled that my parents let me have my party there. It's something I will probably always remember.

-Age 5, Louisville, Kentucky, USA

A neighbor boy that I played with told me he was having a birthday party. I can still clearly remember the day because he was wearing shorts, no shirt, and cowboy boots, and even at that age, I thought he looked very silly. We were eating Popsicles my grandmother had given us. Mine was red and I showed him my tongue. His was purple. When he told me about the party, I felt really excited because in our neighborhood, we rarely had parties. What happened after that comes in flashes. I told my mother about the party being the next day and I remember she griped about having to run out to buy a present, but she did it. I remember getting prettied up and walking down the road to his house, knocking on the door, and his mother coming out. She looked confused. I told her I was there for the birthday party, but she laughed and told me it wasn't a children's party; it was a family party. She didn't even take the present—she just laughed at me and shut the door. I remember crying all the way back home, not because I wanted cake and to have fun. Maybe a little bit, but mostly because I was embarrassed and ashamed of my mistake. This memory is very clear and still oddly painful over 20 years later.

-Age 6, Memphis, Tennessee, USA

My earliest childhood memory is of my fourth birthday. I had a bunny cake that my mom and dad had either made or bought for me. It was really cute and it had the number four as a candle. When I blew out the candle, my mom took it out of the cake and put it on the sink. I remember sneaking into the kitchen when no one was looking and taking a bite out of it, only to realize it was not candy! I had wax coated all over my tongue and I was trying to rinse it off with water. I guess that always stuck in my memory because I was either really disappointed or embarrassed that it wasn't what I thought it was and that maybe I was going to get in trouble for it. I remember it so vividly, still to this day, and now, of course, I just laugh whenever I see a numbered candle for someone's birthday.

-Age 4, Fountain Valley, California, USA

My earliest childhood memory is waking up on Christmas morning. I really wanted some dolls, so I was really excited to open up my presents. I remember my mom making breakfast and I was just so happy to have my dolls that I wanted. I think it's because Christmas time is such a memorable day in our lives, especially when we are children.

-Age 3, Louisiana, USA

My earliest memory would have to be being in Toronto, Ontario over Christmas when I was probably three or four. I remember my mom telling me I was sleeping like an angel. The relatives we were staying with felt bad that I had no gifts to open and bought me a stuffed Snoopy.

-Age 3, Toronto, Canada

It was when I got my first bike from Santa Claus, which my parents asked him to deliver. He was a neighbor of ours who worked in a circus, and he ended up planning this surprise for me. Because it was something I asked my parents for a lot, it ended up staying with me in my memory because I was so happy that day, I couldn't forget it.

-Age 6, Cascavel, Paraná, Brazil

My experience of a Thanksgiving day: the sweet pumpkin pie; the tart, fruity dessert, and, obviously, the rich, chocolate frozen yogurt finished off the stunning supper. I looked from one face to another, secretly expressing gratitude toward God for every relative. There were grins all over the place, from my sibling's devilish smile to my grandma's inspiring grin. I checked out the room with the adornments I had helped my mom set up. Pictures of pumpkins, still up from last Halloween, hung close to Christmas wreaths and holly. In the focal point of the table sat a spilling-over cornucopia, loaded down with wheat, grain, and a wide range of products from the earth.

-Age 11, USA

One of the earliest memories I have involves climbing the refrigerator to get to the cupcakes at the top. I was a pretty short child, so I had to jump to reach the handles. From there, I put my feet against the doors so I could shuffle upwards. I did that until I reached the groove for the water and ice dispensers. That solid footing helped me jump up to reach the top of the fridge. I was able to pull myself up to the top and grab the cupcakes before going back down. Unfortunately, the cupcakes were old. The sheer disappointment of doing all that work for old cupcakes made a deep impression on me.

-Age 4, Los Angeles, California, USA

My mom made Christmas cookies but, as a joke, she put these hard little beads on the cookie that were very hard to chew, so I was sad that I couldn't eat the cookie.

-Age 2, Abrams, Wisconsin, USA

My earliest memory is of my first birthday and the cake that my parents made for this occasion. I was obviously very young, so the cake was mostly jelly and cream, with some Maltesers and Smarties embedded in it. I was walking, but not great at it. I remember the blue dye from some of the Smarties (European ones, they are more like M&Ms, with a colored chocolate shell) getting streaked through the cream as the cake collapsed a bit. After they sang happy birthday to me and celebrated a bit, which I didn't really care about, I was let at the cake, and I enjoyed taking big chunks out of it and eating the jelly inside—but I was aware I was under close supervision. I remember that it was just my parents there, and the two family dogs were not allowed into the sitting room where this was taking place for cake-related reasons. I think this stuck with me because it was a special occasion, and I was not normally allowed that much sugar at that age. I was not used to celebrations like that, no matter how small.

-Age 1, Dublin, Ireland

When I lived in Texas, our next-door neighbors had a boy my age named Andrew. Andrew and I were decently good friends, and before either of us started school, we would often play in either one of our backyards in the afternoons. When I was around three, Andrew pushed me, and I fell backward and into his sand pit. I wasn't hurt at all—I just remember being irritated that I now had sand all in my hair. Playtime was over. I got up, yelled at him—something along the lines of, "No pushing!" and stalked back home. My grandparents were in town at the time, and they asked me why I was home early. With all my three-year-old might, I spat, "Andrew push!" They were shocked for a brief second, then my grandmother knelt down and asked me, "Why did Andrew push you?" I thought for a hard second. Why had Andrew pushed me again? Oh, right! "I want Andrew to eat sand," I replied matter-of-factly. So, I tried to shove sand in poor Andrew's mouth, and he retaliated by pushing me. My grandparents had a good laugh, and I was back at Andrew's within the next half hour.

-Age 3, Texas, USA

My earliest memory is playing at the playground with my friends. I was roughly five years old. All the kids from the neighbourhood gathered and we played hide and seek. It was a warm summer day, and I enjoyed playing with the other kids. I believe there were about 15 of us, just playing the whole afternoon until the night fell and the streetlamps lit up. I remember my mum calling me from the balcony to come back home for dinner. I was annoyed at her because some of the older kids were allowed to stay out longer, and I wanted to keep playing. I had to listen to her, so I picked up my younger brother and headed home for dinner. It was a day full of joy and happiness.

-Age 5, Location unknown

My parents were getting a new dog and she had to come from Texas because of the breed. The people drove up to us in New York, but they got delayed by a couple of weeks. So, when they finally arrived, we named the dog Godot, like the play *Waiting for*

Godot. I remember the van pulling in and the couple and the dog getting out and being so excited. It was an exciting day.

-Age 3, Buffalo, New York, USA

My earliest childhood memory is from the day when my daddy brought a dog into our home. It was memorable because I wanted a puppy for so long. My sister and I named him Dark because his fur was black like the night sky and his paws were brown. He was shy at first; he looked scared. I remember that it happened in spring on a beautiful, sunny day. It was a surprise and it filled my heart with joy. I still remember his funny nose and scared eyes. I cuddled him a lot. I let him smell my hands to get to know me and he felt reassured by my touch on his tiny head.

-Age 6, Rome, Italy

I remember that when I was 10 years old, my parents woke me up in the morning and presented a dog as my birthday gift. It was a wonderful day in my life. It is one of the best gifts I've ever gotten for my birthday. I named my dog Neo. I spent so much time with my dog. These times were my golden times. I will never forget this memory.

-Age 10, Dayton, Ohio, USA

I remember checking the screen door to see if it was locked, as I often did as a toddler. Once, to my surprise, it was unlocked! I looked around behind me to see if my mom saw... nope! I was FREE! Carefully, I went out the door, and let it shut quietly behind me. Which way should I go on my adventure? Behind the fence where the condos are! So, barefoot, in only my diaper, I walked on the sidewalk to the end of the street, turned left, and went to the condos with the huge grassy area to play. Immediately, a huge dog ran up to me, barking and drooling, looking me straight in the eye, so close I got spit on as he barked and I could feel his breath. Terrified, I turned and ran home, feeling the hard, rough concrete as my heels struck, wishing I had my diaper-holding hand free to pump for faster running. I expected a bite to the neck at any moment. As I ran, I glanced over my shoulder to see if he was still there. Yep, hot on my tail. I dashed into the house, slamming the door behind me. What an excellent adventure.

-Age 3, Lakewood, Colorado, USA

I was awakened on a rainy night after being put to bed for the night in our apartment in Chicago. I was two and a half. My grandmother carried me to the couch and got me dressed. I remember her putting on my socks and me asking where we are going. She told me we were moving to a new house. I remember this because it was my first and last memory of living in Chicago. I remember the rain and a long drive.

-Age 2, Chicago, Illinois, USA

I lived with my parents, my older brother, and my younger sister in a little village and I had a dear friend. I remember that my father needed to move for his work and all the family went with him. We left the little village for the big town. I felt totally crushed by the events; I remember myself watching through the back window of the car, searching for my friend... I remember the tears and the fear.

-Age 5, Tourette Levens, France

One of the earliest and more profound memories of my childhood is of stockpiling pieces of different gum during my last year living in Eastern Europe. I was planning to eat all the gum on the train as it went from our city to Moscow to board a plane to the US. I didn't succeed and stopped after stuffing my face with about four of them. I really liked chewing gum, and this was one of the ways I would save up for a last hoorah. It was a profound memory because it highlights that I was so little and my mind so narrow, yet I can remember how sad and difficult it was for the rest of my family. My parents and grandparents all were together, but everyone had a different thought in their minds. Yet all I cared about was chewing the gum.

-Age 8, Ukraine

My earliest memory is about the day we moved out of our first home. I remember my parents carrying boxes to a truck while I was inside. I kept running through the empty house and finding hiding places. I remember sitting in a closet that had a little space where my parents couldn't find me. They kept yelling for me as I sat in the closet. Finally, I crawled out of the space, crying. I remember crying because I heard my parents yelling for me and I was scared. It is the only memory I have of living in that house.

-Age 2, Flint, Michigan, USA

I remember being at home during an earthquake. I was playing video games with my brother and I heard a rattle across the floor. While my brother and I bunkered down for earthquake safety, we kept hearing the rattling go across the wood floor. Through the doorway we could finally see what made the noise. It was a spoon that somehow fell off the table and slid all the way down the hallway.

-Age 5, Long Beach, California, USA

I was living with my family in South Sudan, in a big hut that was our home. We were used to heavy rains and lightning strikes during the rainy season, but one day, it was particularly intense. The rain was pouring down in sheets and the lightning was flashing all around us. Suddenly, there was a loud crack of thunder and a bright bolt of lightning struck our hut. I remember feeling a jolt of fear and confusion as I saw flames start to spread throughout our home. The heat was intense, and the smoke was choking. My family tried to salvage what we could from our burning home, but it was a losing battle. Everyone was devastated as we watched our beloved hut burn to the ground. The loss of our home was a huge blow, and we struggled to pick up the pieces in the aftermath of the fire. To this day, the memory of that night is etched in my mind, as a symbol of the power of nature and the fragility of our lives.

-Age 7, South Sudan

I was sitting in the laundry room on the floor and I was all by myself. I could smell the detergent as I was peeling hard bits of it off the detergent bottle using my nails. I enjoyed the feeling of it. I remember that the detergent was blue and the room was dimly lit. I thought I was doing something naughty and getting away with it.

-Age 3, Vanderbijlpark, Gauteng, South Africa

When I was about five years old, I went on a family vacation with my grandmother and great-grandmother. I don't remember much of the trip, but I remember feeding ducks on the last day we were there.

-Age 5, Williamsburg, Virginia, USA

My earliest memory was being in the car while traveling cities with my parents and passing by a bike shop. It was on the side of the road and had pink walls. I remembered asking my parents to go in and look at bikes. We went inside and had a quick look. There weren't many bikes for my age, but I remember one green bike that I really liked. I don't remember as many details from inside the shop, but I remember we didn't get the bike and I was slightly upset. Once we left the store, we got back in the car and continued the trip.

-Age 3, Albacete, Spain

When I was a toddler, my family was at Walt Disney World for a trip. I was riding on one of the attractions, The Barnstormer, with my father. It was dark outside, and the line for the ride was fairly long, even by Disney World standards. The ride was like a rollercoaster but for little children that can't handle the larger rides just yet. It was an enjoyable time and had me feeling pure excitement. Especially after standing in line for so long, it was relieving to finally get to experience the attraction. Hearing the noises of the chains pulling us up the relatively small hill, and the sound of wind blowing into my face as we went down. The ride was a tad short, though, and it left me disappointed that I had more lines to look forward to that night. I wished the ride didn't have to stop, and I could just enjoy those feelings for as long as I felt like.

-Age 3, Orlando, Florida, USA

When I think about childhood, my grandmother's house always comes to mind. Enohe was her name, now dead. Magical place. The delicacies she made, the delicious food for lunch. The place was simple and very well cleaned and cared for, the plants everywhere, as well as the small spice plantations. This memory is a constant: the color of the yellow house, the wall very close to the street where we played a lot without worrying about cars, childhood life in its essence. I remember because it refers to affection, zeal, and care; it refers to the love of the family.

-Age 8, Juiz de Fora, Minas Gerais, Brazil

I remember being at my grandmother's house in Michigan. My grandpa made a kind of teepee of sticks against a tree in the yard. I used that as a sort of fort to play in. I had good times visiting my grandparents. My grandma made her own bread, which she rose in the attic. I can still smell the bread baking. I guess this stands out because it was so perfect.

-Age 5, Lansing, Michigan, USA

The family had planned a Sunday morning outing to the famous Rose Garden. This memory was probably my first encounter with a possible accident. I remember running around the garden away from the watchful eye of my parents. I was running as fast as I could for no apparent reason on a concrete path when I encountered an open manhole with a deep trench to the drain. I clearly remember I stopped right at the edge of the opening, almost leaning over into the hole, looking at the reflection of my petrified face over the water in the drain. I can still feel my heart pounding to this day. I always wondered, "What if I had fallen into the drain…?" Would it have been deep enough for me to drown, or could anyone hear me if I was trapped inside?

-Age 6, Chandigarh, India

My earliest memory is of breaking my front teeth out. The memory isn't perfectly clear in my mind, mostly just a collection of images. I remember that my parents were rearranging furniture, and as a result, most of the house was vacant while things were stacked up in one room or another. I was running around the house as fast as my little toddler legs would carry me, having fun. I went into a room that had a big plush rug, at the center of which was a glass-top coffee table. My toes caught under the rug because, of course, I was barefoot. I went flying face-first into the glass-top coffee table and busted my face right through the glass. Fortunately, I didn't sustain any major injuries other than losing my front two baby teeth very early. I had a large gash on my lip that required stitches, and I have an almost imperceptible scar now as a result. I don't really remember the pain or being scared, but I definitely remember the exhilaration of doing something I normally would never have been able to get away with, running in the house.

-Age 3, Vinton, Virginia, USA

I remember I was very small. My cousin, five years older than me, was playing with his friends on the floor. At one point, when I approached and lowered my head to look at them, one of my cousin's friends, without seeing me, got up and hit his head on my nose and I felt a very strong pain. I believe my nose might have broken because to this day, I have a higher part of the bridge of my nose.

-Age 4, São Paulo, Brazil

One of my earliest memories would be when I was age four to five and was up at my cousin's house. He had this rubber-covered metal hammer (kind of like a toy hammer, but still solid). He was throwing it up into the trees to try and knock loose some mushrooms that were growing on the trees above us. I don't remember much of the time leading up to the "incident," but basically, he threw the hammer up, it bounced off a tree and came straight down on top of my head. All I really remember visually from the incident is a specific moment where I was crying, holding my head, and walking into my cousin's garage, and then about two to three hours later, being in the hospital after getting stitches.

-Age 4, Montpelier, Vermont, USA

My earliest memory is having my forehead stitched after jumping off a table and busting my head open. I remember the cloth over my face with the opening over the wound and seeing the bright lights of the exam room through the cloth. I didn't learn my lesson because I did it again about a month later and had to be stitched in the same place again. I have a nasty scar from it and have to wear bangs to this day.

-Age 3, St. Clairsville, Ohio, USA

I remember breaking a glass on my arm when I was four, and my mom started to walk me to the hospital, which was several miles away, because my father had our only car. My grandfather, who lived in the same neighborhood, was driving by and saw us, and picked us up. I had to get stitches.

-Age 4, Ohio, USA

I feel as certain as I could possibly be that the first memories I remember are solely my own and not based on family stories or photos. The visual and emotional memories have been vivid my whole life. I was hospitalized for whooping cough, now preventable by vaccination. I remember being in a crib with very high sides in a ward where cribs and beds were separated by curtains. What I remember most was how I was treated by the nurses because most were kind, but one was very gruff and rigid. I recall feeling very alone, missing my parents, and wanting to go home. My dad brought a green plaid cot and put it in the back of the car to take me home. It is interesting to me that I have returned to those memories many times during my life because I like knowing that I can remember them so well. I don't know why that is important to me, but it is.

-Age 3, Kingston, New York, USA

This instance occurred in the backyard of the first house I remember living in, during a time of nice weather. The unpleasant experience was on a morning when there was a heavy dew on the grass, and I picked up the end of an extension cord that was plugged in, the end lying on the wet grass. The shock sensation was immediate and intense. I let go of the end of the cord and ran crying into the back door of the house to my mom, who was always there. That is the extent of the memory, though I have a very crisp visual picture of the scene, the green grass and the white house and garage, and the small, concrete back porch with steps.

-Age 4, Pampa, Texas, USA

My earliest memory is of a vacation to Canada right before Christmas. I remember going to ski school with an instructor and several other kids my age. I remember that myself and three other kids were moving at a faster pace than the rest of the kids. The instructor took just the four of us after lunch and said were going on a ride. We went into this area where a big bus on a cable came in and people got out. All four of us got into it and sat down and waited for all the other people who were way older than us to get on. The doors closed and we started moving. We slowly went up the mountain and I remember feeling all weird for a minute and then wiping my nose. The instructor looked concerned and told me my nose was bleeding and got a tissue from another person on the tram. I don't recall much after that distinct moment. I think it scared me because people looked worried that my nose was bleeding, and I was so little at the time.

-Age 3, Whistler-Blackcomb Resort, Vancouver, Canada

We lived in a two-story house that had a walkway at the top of the stairs that led to the actual space on the second floor. On that walkway, the railing had vertical bars to keep you from falling off. I was pretty young and just learning to get up the stairs all by myself. But I was also a curious child that liked to get into trouble. So, I looked down through the bars because it was cool when you're small. Well, I stuck my head between them and got stuck. I'm sure I cried a lot. I don't remember how I got unstuck, personally, but I've been told my dad had to cut one of the bars out to free me.

-Age 3, Silverdale, Washington, USA

I am about three and I am sitting on my mother's lap. We are in the bathroom and she is sitting on the closed toilet. She is wiping my forehead because it is bleeding. She is fussing and saying I was lucky I didn't kill myself. I remember feeling safe and secure. My dad walks in and gives me a strange look and asks me what in the world I was thinking. It followed an event that I don't remember where I accidentally locked myself in a room and panicked and threw myself through a second-story window and was hanging there before my father was able to rescue me.

-Age 3, London, England

My brother had a turtle that was kept in his room. I remember the glass bowl and the round sort of beveled shape of the bowl

53 | EARLIEST MEMORIES

and the texture of the gravel in the bowl. It smelled a little bit. I remember knowing I was not allowed to play with it. But I desperately wanted to play with it. I don't remember taking it from my brother's room, and it is murky to me where I ended up taking the turtle. In my memory, I took it to the street outside my house where there was a puddle because I wanted to see the turtle swim. Definitely, there was a puddle, with muddy water in it. My mother must have noticed me missing because she came to find me. I don't remember her calling to me but as she drew near, I remember that terrible, anxious feeling of knowing I had done something I was asked not to do—compliance was expected in our house! So, I picked up the turtle and put it in my mouth, I guess to hide it. That part I only vaguely remember. I don't remember the feeling of a turtle in my mouth. The picture of my mother seeming so tall to me, leaning down towards me to find out what was going on, is very strong. I think I burst into tears.

-Age 3, Saugerties, NY, USA

My uncle had come to visit and was going to spend a few days with us. He had just come back from a long trip in Norway that had lasted a few months, so it had been a while since I had last seen him. He brought back gifts. I don't remember what he gave to my older brother or my parents, but he gave me a toy: a white pony made of plastic with a rainbow-colored mane and tail. It was fascinating—I had never seen anything like that before! I wouldn't let go of it all day long and would bring it everywhere with me. I'm not sure why but, at some point, I felt like I wanted to cut the pony's mane, just a tiny little bit. But my mind was screaming that it was wrong and that I shouldn't. I felt like opposite thoughts were colliding, but I couldn't stop myself from doing it, and, ultimately, ended up taking a pair of scissors for children and hiding behind my bedroom door. I knew I shouldn't do it and that was why I was hiding. *"I should stop. Mom is going to be angry. The pony won't be as pretty anymore... but just a little, maybe? Just a tiny little bit...?"* I ended up cutting a little part closest to the pony's forehead. Most of the mane was still untouched, but I felt a wave of guilt washing over me, and regret too. I felt I had done something foolish and damaged something I liked with my own hands, for no reason, and felt sad and resigned, but never cried.

-Age 6, Casablanca, Morocco

My mom and I were waiting for my father. He was in the Army at the time, and we were standing by a building. I'm not sure what the building was but I'm sure it was at an airport because I remember hearing the jets. I wasn't really feeling any emotions that I remember, except for curiosity. I assume this one stuck with me because I was getting to meet my dad after a presumably long absence. The memory is more of a snapshot in time than it is a complete memory.

-Age 3, North Carolina, USA

I remember being sick with the stomach flu, a common occurrence for me. I can remember lying in bed, feeling as if my stomach was swimming in an ocean, with the current ebbing with each breath. There was a standing fan blowing cool air on me; the noise of it was a comfort. My sheets were purple with pink flowers. Our family dog, Pepper, was laying at my feet, a comforting weight and presence. I can recall looking up at the ceiling, and counting the plastic glow-in-the-dark stars in my head: "One, two, three..." There was a plastic glow-in-the-dark moon in the shape of a crescent, and the bottom half of the moon was peeling off the ceiling. It looked as if the moon was going to fall. That bothered me most of all, more than my queasy stomach churning like ocean waves: the fact that the plastic moon was going to fall on the ground, and I wouldn't be able to catch it. As I look back on that memory, I feel as if it is incredibly silly. My worries were inconsequential and small. But to my four-year-old mind, the worry of my moon falling off my ceiling was my biggest worry, so much so that it left an imprint in my memory.

-Age 4, Greenville, South Carolina, USA

I was in bed, sick from a cold. I was snuggled in my parents' bed since they were out for work. The only other person I remember seeing was my uncle who was babysitting me. I think that I was scared about being sick and my uncle said something to me that comforted me.

-Age 6, Chicago, Illinois, USA

I went to my Granny's place to spend the Easter vacation along with other cousins. My parents didn't go; I was alone, apart from my family. I had great fun playing with my cousins but then at night when my mom called, after hearing her voice, I started weeping. I didn't understand why, as I had a great day, but I couldn't stop. Seeing me, my cousins started to laugh and made fun of me. I hung up the phone and ran to bed and didn't talk to anyone. The next morning, my mom came and I insisted to leave with her and didn't even greet my cousins.

-Age 5, Ronkonkoma, New York, USA

I was inside an apartment, alone near the front door. There was a metal toy airplane that had blinking lights when it was pushed along the floor. It smelled of an electrical motor getting hot. I was curious as to how the lights worked without a battery. I examined the plane for a way to take it apart. After noticing screws, I went in search of a screwdriver. After finding one, I attempted to take it apart. I remember being very determined. However, being unsuccessful, I wanted to know why I couldn't remove the screws. I discovered that the screwdriver wasn't the right type. This is when I learned about the two different types of screwdrivers. 'Til this day, there is nothing better than taking something apart to see how it works.

-Age 3, Georgia, USA

My earliest childhood memory is of when I went to Colombia to meet my father's family. I remember the travel was boring because of waiting in the airport. When we arrived, we went in a Jeep and traveled through the city into a rural area. It was full of mosquitoes and flying insects; I hated it. This was also a long ground journey because our destination was in the middle of nowhere. When we arrived, I was all itchy and wanted to sleep, but my parents wanted to party with the others and turned on music and danced all night long. I spent two days doing nothing, just playing with my toys in my room, until we finally left.

-Age 7, Colombia

The earliest memory that I can recall is of being behind my garage. I was sitting outside, there was grass everywhere, and there were a bunch of butterflies that landed on me. I remember my mom telling me not to move because she wanted to get a camera. She ended up getting a camera and taking pictures, but I don't remember what they looked like. It was the first time I remember being connected to or part of nature. It was amazing to have these little winged creatures land on me. It hasn't ever happened since, and I think that's why it stuck with me. I always wanted to have it happen again.

-Age 5, Virginia Beach, Virginia, USA

My earliest memory is when I was sitting in the doorway at my old house. The doorway connected the kitchen to the hallway. I remember having dark, wooden floors. I had my favorite toys at the time—ponies! I specifically remember having a wild imagination. I really went into my own world when I played with my toys. Each pony had a unique personality. I cherish that memory. It reminds me of my innocence and my ability to be happy with simple things.

-Age 4, Jasper, Indiana, USA

I was fascinated by the newly launched plastic crayons my dad got me. I was always a kid who was captivated by different colors and stationery items but could never keep even a paintbrush. I was so attached to this box of crayons that I always carried it with me. I showed it off to my other classmates. They were jealous of tiny me. One day, I carried it in my school blazer's pocket to the toilet. While I was wearing my pants, the box somehow fell into the commode. My heart froze. I immediately called the caretakers. I pointed to my swimming box of crayons and started crying. My class teacher arrived at the "crime scene." She explained to me how some things are meant to leave and one should let it go. She took my tiny hands in hers and made me flush it down. I squinted my eyes doing it. My heart sank, as my box did. Later, my dad bought me an even bigger box of plastic crayons. It had 36 shades instead of 12. That's why I love my dad. Also, this made me learn that some things are never meant to stay with you. You can cry, make a fuss about it, but eventually will come out of it in peace.

-Age 5, Grand Junction, Colorado, USA

My earliest childhood memory is getting my green light taken from me on the school bus. In elementary school, the teacher used to give us a reward for being good in class. The green light (like the traffic light), was given to us so that we could take it home to our parents. Well, while I was on the bus, I was showing my friend my green light. A small boy near me liked it and told his older sister that he wanted one. So, when they were getting off the bus, she snatched it from my hands and gave it to her brother. I was crying and I was very angry. I told the bus driver, but there was nothing that he could do about it. I just wanted to punch her in the face. I still do. I think I just felt really hopeless. I wanted to show my parents that I had a good day, and it felt like she took that away from me. I hate her. I mean it when I say I hate her.

-Age 7, Mt. Clemens, Michigan, USA

I was standing on top of the jungle gym at my preschool with a friend, about to head down the slide. It was cold outside because

I was wearing my big, puffy, colorful coat. I could see the whole playground from up there. I'm not sure why that moment stuck… maybe I felt powerful. I was high up. I loved that jacket because I thought the poofiness made me look muscular.

-Age 5, Greenbelt, Maryland, USA

My first childhood memory was when my class in elementary school went to a forest nearby for the first time. It was such a great summer day. I remember the sun, all the noises of birds, and the smell of leaves and grass in the forest back then. We were there for a couple of hours and our teacher showed us everything she taught us about in school so we could relate better. The highlight of this day was a squirrel that ran on the trees there making funny moves, so we all laughed so much.

-Age 7, Thuringian Forest, Germany

The first specific recollection I had was in late 1951 when we had moved back to Colorado. I remember going into my grandpa's bedroom and crawling into bed with him. What sticks in my mind is that he wore a long nightshirt and no other men in the family wore such a garment.

-Age 2, La Veta, Colorado, USA

My earliest childhood memory is of my sister teaching me to do a cartwheel. I remember being in the den of our house and her teaching me how to do it. She had just learned how to do cartwheels in cheerleading and she was doing them in the house. She showed me how to do them and I tried to emulate her with poor results. I remember falling and never fully committing to a full cartwheel. In the end, my mom told me to stop or I would hurt myself.

-Age 3, Clemmons, North Carolina, USA

My earliest memory is going to a pool with my dad, sister, and brother. The pool was at my dad's friend's house. Dad's friend had two sons who were also there. The pool was outdoors, attached to the back of their house. I remember, in particular, after arriving and getting out of the car, going up onto the diving board and jumping into my father's arms as he was standing in the pool. I was too young to swim at that age, so I needed him to catch me. I was very excited to jump into my father's arms from the diving board; it was a new experience. I remember the smell of chlorine, the feel of the heat, and the water cooling off the heat.

-Age 4, Leominster, Massachusetts, USA

I was barely able to walk and I was with my mother and a friend at her friend's pool. I stood on the deck, walked towards the pool, and dove in. I felt myself surrounded by the cool water and watched the sparkling light from the sun dance on the water and the waves that I had just created. I remember sinking all the way to the bottom and not being afraid at all. The world around me was peaceful and felt calm and I had no idea of the situation I was in. Soon, I began to struggle and thought of needing to breathe, but continued to just breathe out when I needed to. I had a thought that I was in trouble now but didn't know just how bad it was for me. Then, I saw above me the form of my mother diving into the pool. She had really long brown hair and it was flowing everywhere around her. She swam down to me and had a scary look on her face. She grabbed me tightly around one arm and I was jerked forward to the surface. We made it to the top and she picked me out of the pool and placed me on the patio with the words, "Stay here!" She got out of the pool and hugged and kissed me. Once I was dried off, she told me what I had done. I had no idea that she had saved my life until many years later.

-Age 2, LaPorte, Indiana, USA

I remember jumping off a dock at my great uncle's house and into the water. All of the adults were, of course, upset and working to pull me out, but I couldn't understand what the fuss was.

-Age 2, Armada, Michigan, USA

My earliest memory is of my aunt. She was probably in her 20s and still living in the same house with my grandmother. I recall one day my mother took me to my grandmother's house for my aunt to take me to a local public pool. I do not recall the actual travel to the pool, but I do recall being at the pool and feeling shy. I was shy about wearing just a small pair of children's swimming shorts. However, I really loved bathing in the pool with my aunt. I recall staying at the pool until sundown and when we left, it was already dark outside.

-Age 3, Suceava, Romania

My earliest childhood memory is with my aunt as she used to babysit me. I recall her giving me a warm bath, dressing me for school, and preparing my lunch box. My aunt used to give me my favourite chocolate wafer biscuits every time she dressed me for school. To this day, I sometimes tell her that it was one of the fondest memories I had growing up.

-Age 4, Kaduna, Nigeria

I remember being on a passenger ferry to France on a family holiday. I must have been three or four. We were in a small cabin, and there was me, my parents, and my older brother. I think the sea was quite rough at the time, as I remember the ferry moving around a lot. My mum gave me a packet of crisps, probably to distract me, and this is what stuck in my mind. I vividly remember the crisps, the brand, which was the UK supermarket Sainsbury's, and the colour of the packet, which was orange. The logo was in white. I think they were chicken-flavour crisps.

-Age 3, English Channel, UK

I was sitting out on my great grandma's back porch as the summer heat hit the back of my head. The birds in her birdhouse were flying around and the Oklahoma wind was hitting the wind chimes. She got flour and salt and water and had me come in to make homemade play dough. She died at 99 years old and was my best friend since I was born until that day. I think of this memory often. I have severe anxiety and I think of this memory to calm me. I remember this because life was good; it was pure. I was a child and my imagination was alive. I see life unscared, unbroken, and my best friend is still here. When I feel conflicted, I think back and I feel I'm almost there again.

-Age 4, Clinton, Oklahoma, USA

My earliest childhood memory is from when I was in pre-K. It was nap time and I was laying down on my mat. I remember seeing a bee flying around. It then landed on my leg and stung me on my thigh. I cried out in pain and got an ice pack for it. It got pretty swollen and hurt badly. I was afraid of nap time at school after this… and bees.

-Age 4, Batavia, New York, USA

I was standing in the yard by the hydrangea bush that used to be there, in the twilight. It was not yet dark, but it was around the time that they put me to bed. It felt like such an indignity, to be forced to not perceive anything during sleep, as well as how they treated me ordinarily—like my thoughts and actions had no value, like I could not control my life, like I was worthless. I remember thinking—with startling clarity—that I was a child, but I was a functioning, sentient, intelligent human, and I was just as aware as they were, even though they treated me in an unfair way and said that children should be seen and not heard. It ignited a spark of defiance that has stayed with me and influenced a number of my actions over the last 20 years. It was the formative moment for me.

-Age 4, Raleigh, North Carolina, USA

My earliest memory involves going to a friend's house for a play date. This was my first play date without a parent. The house was in a great part of town and had a mini-carousel in the backyard. I remember wearing a really pretty party dress for the play date. We had lunch together, though I was not comfortable being there alone, and I had an anxiety attack and started crying. I was not able to console myself, so my mom ended up picking me up.

-Age 5, Providence, Rhode Island, USA

I was in the sandbox in my backyard. It was a sunny day and I was enjoying playing. I knew my mother was going to be coming out the back door any second to check on me. It was the first time I really became aware of what I was doing.

-Age 2, La Porte, Texas, USA

My earliest memory is of going to the zoo with my mother and stepfather. It's getting cloudier as I grow older, but I can still remember being carried on my stepfather's back at the gorilla exhibit. I remember wearing a yellow t-shirt. It's a pleasant memory, but I would say that it has changed in the way that I don't view it from the first-person point of view anymore; it's more or less me watching myself from the third person. I didn't have a very pleasant childhood, so I would say I subconsciously tried to hold on to a memory where my family got along for once.

-Age 3, Washington, D.C., USA

I was in my living room on a couch and my family and their friends were sitting down, too, and watching TV. It was bright from the lights. They were passing around a joint. My aunt passed it to my uncle, but he wasn't paying attention, so I grabbed it and pretended to puff on it. I only pretended because I knew that kids shouldn't smoke. I felt like it was funny. Everybody made concerned noises and I passed it back. I remember everybody making that concerned sound, breathing in really fast, like "ahhhhh." I think that the noise of concern was kind of a jolt.

-Age 3, Levittown, Pennsylvania, USA

The earliest childhood memory I have is of being a toddler, perhaps about one and a half to two years old. It was a Saturday and my mother went to visit her mom (my grandmother) who was a street away from us. I was placed down on my grandmother's kitchen table, laying on my back and looking up at the ceiling with everyone coming to see and play with me. Everyone includes my grandmother, aunt, and older cousins. I recall the atmosphere being very serene and everyone was happy. I remember clearly it was early in the morning. My mind was also wandering and I felt very intrigued. I feel like this is the first time my mind was able to comprehend my surroundings and I was able to recognize people, thus it's a memory that has stuck 'til now.

-Age 1, Location unknown

When I was really young, I remember being at church with my family. I was a very shy kid and, even into my teens, didn't really talk to people. So, during church, I mainly kept to myself, coloring books, being quietly rambunctious, the normal things kids do in church. At one point though, I remember glancing back at the pew behind me and seeing this girl. She was probably 15 or 16—so basically an adult to a five-year-old—and I just remember not being able to take my eyes off her. I stared and stared, probably for only a few seconds but what felt like a week. Finally, she noticed I was staring and gave me the sweetest, warmest smile. I was so bashful I immediately turned away, horrified. But I couldn't resist, and every once in a while, would slowly and quietly turn around to sneak another peek at this girl, secretly hoping she would notice and smile at me again. And every time she noticed and would give me that smile. I was, in my five-year-old brain, in love. I have no idea what her name was, or who she was, but I'll always remember that smile she gave me.

-Age 5, Mississippi, USA

When I was 10 years old, I had a good relationship with a boy of the same age. He spent most of the time with me every day. He studied with me, played with me, and took care of me when we were alone. He understood me a lot more than others and I can still feel how he looked into my eyes whenever we were together. That was my most special and unforgettable memory. It was my first relationship with someone of the opposite gender and it was a happy moment.

-Age 10, California, USA

My earliest memory is when I was about two years old, at some type of family gathering in the backyard of a relative. I remember I was trying to walk down the stairs from the back verandah to the grassy area, but I was clinging onto the railing and too scared to walk down the concrete steps because I thought I was going to fall. I remember the feeling of holding the metal railing really tightly. It felt cold, and I think it was painted dark red or brown.

I remember hearing my dad calling out my name, as he was sitting a few metres away on the side of the grass. There were lots of people around, lots of conversations, and just general party noise, but I distinctly remember hearing my dad call out to me and looking over to him and feeling less scared because I knew where I needed to go. He was calling me over and I eventually made it down the stairs with his encouragement and walked across the grass to my dad. He showed me a small fish pond in the backyard and I was excited to see the fish.

-Age 2, Adelaide, Australia

I remember my grandmother made a dress for me that had kittens on it. I vividly remember trying the dress on for the first time and looking down and seeing the kittens. The dress was a blue print with big kittens on it with blue eyes, and big pockets on the dress. I remember being so happy and feeling so special in my kitten dress. I remember my grandmother sitting with me in her lap talking about the dress and my grandpa making a big deal about how pretty I looked in the dress. I also remember her taking me to Sunday school and church in that dress. I remember feeling very loved and special, a very positive, warm memory.

-Age 3, Gulfport, Mississippi, USA

My earliest childhood memory was when I was four years old. I was in a school carnival and I was dressed as Fred Flintstone. I was accompanied by my cousin and I remember having a great time, eating a lot of candies, and walking in a big square. I think it was a moment where I became aware of my surroundings and

I told myself that I would remember that moment forever.

-Age 4, Miami, Florida, USA

My earliest memory comes from some point while I was in preschool at Bridges to Learning. I recall foot-racing my dad to the car from the school and it being cold and dark outside. So, this was probably fall/winter, making me around three and a half years old. It wasn't all that far, maybe 40 yards, and it wasn't a parking lot—the car was at a grassy curb. I remember we did this regularly. He always made it close, but he never let me win.

-Age 3, Rockaway, New Jersey, USA

On my first day of school, I got into trouble a bit. First, I flushed a urinal repeatedly because I was irrationally angry that someone else flushed it. Second, I squeezed the juice out of an orange because I didn't like oranges, but did like orange juice, and it made a mess. I remember having to sit in a corner. Then, on the bus, I misunderstood where my stop was. I thought that the bus would go straight to my door. I almost stayed on too long, but my parents noticed and came to get me.

Age 5, Albany, Georgia, USA

The most stressful memory I remember was when I was just starting kindergarten. My mother sent me to school wearing my brother's underwear (I suspect because I didn't have any clean ones—at the time, there were six of us living at home and my mother had her hands full!). When it came time to play Duck, Duck, Goose, I refused to play because I was afraid my dress would fly up and everyone would see my brother's underpants! How humiliating for a five-year-old could that be?

-Age 5, Detroit, Michigan, USA

I attended Christ Church Preschool when I was three and four years old. Several of the kids in the class are people I still know today. All of us were lined up outside. I remember eating dirt as part of some way to impress the other children, like, "I'm not afraid to do that!" I think I rehearsed this memory a lot and told people about it routinely—who the hell knows why? This one probably took place in springtime.

-Age 3, Charlotte, North Carolina, USA

I remember being at some kind of daycare facility and these kids were watching TV. I put on a fireman's hat, which had a loud siren on it, and the daycare teacher yelled at me for making noise when the kids were watching TV. I think I remember this because I don't like authority and this is the first time that someone besides my parents told me what to do, so I did not take kindly to it.

-Age 3, Baltimore, Maryland, USA

The first memory I have is the first day of kindergarten and my dad is dropping me off. One of my best friends was already at the school. I remember looking up at her and waving. She looked at me, didn't wave, and began to walk away. I started crying because she wouldn't say hi. I do not remember what happened after that.

-Age 5, Philadelphia, Pennsylvania, USA

I'm waddling forward in a rather ungainly way. I think I'm uncertain. I'm moving toward a small black and white ball and I kick it. The ball flies away and I run after it, but I knock into another child who is also running after the ball. We fall down and cry, but my mom picks me up. I think our parents do a little bit of organizing because the next thing in the memory is me and the other kid trying to kick the ball in each other's direction.

-Age 3, Lancaster, California, USA

My first childhood memory is when I was living with my parents. Going out to the park to play soccer with my father and mother, having a lot of fun with them, and hugging them a lot. I remember that aroma of my parents: so beautiful, an aroma of flowers. Those sunny days when we enjoyed playing in the park on a Sunday... I will never forget my parents' smile and how much fun I had back then.

-Age 8, Los Angeles, California, USA

My earliest childhood memory was when I was three years old. I remember my nanna picking me up from my home and taking me for a walk in the sunshine. I was not in a pram but walking side by side with her as she held my hand. We walked for 30 minutes to our local park. She bought me an ice cream from the snack stand and we walked through the park. It had lots of lovely flower beds that were in bloom and I vividly remember the smell of flowers and roses. There were lots of bees flying around the flowers and I tried to touch one and my nanna pulled me away. We then went into a big greenhouse, and I always remember the smell of flowers and what I now know was fertilizer. It was very hot and humid in the greenhouse, and we left to get some fresh air. After that, we sat and watched the bowling on the greens. I remember the bench, how my little legs felt on the wooden slats, and how old all the players looked in their white uniforms. I finished my ice cream and my nanna pulled out her tissue and wiped my face as I tried to pull away. We then left and walked back home.

-Age 3, Barrow, Cumbria, UK

My earliest childhood memory is being in a pram being pushed by my mother down a street. I can recall looking up—I was laying on my back. I can remember seeing the trees above me swaying in the wind. I saw a canopy of green leaves overhead. I actually still remember which street I was being pushed on when this happened. I don't remember any emotions or feelings, just the visual of seeing the leaves and trees overhead and being aware of exactly where we were. It seems out of nowhere, and I don't really remember much else from the first five to 10 years of life. But that one particular memory is very vivid in my mind. We did later move to a house on that exact street.

-Age 1, Winston-Salem, North Carolina, USA

My earliest memory is of a walk I took with my mother. It had been raining before we took the walk, apparently, as the streets and pavements were quite wet. I remember looking down and seeing a lot of worms crawling around and thinking how icky they were and how I didn't want to step on them. I can remember pointing them out to my mother and then continuing to look down on the sidewalk and stepping over them very carefully. My mother made some kind of comment to the effect that there certainly were a lot of them. I was a bit afraid and intimidated by them, but I stepped over and around them and made it through. I think this may be my earliest memory, due to the fact that it was just totally unexpected and surprising for me, as well as it being quite unsettling.

-Age 4, Denver, Colorado, USA

My earliest memory is of running across our yard in the springtime to the mailbox where the daffodils would bloom. I remember it being sunny, the grass being green, the flowers being yellow, the skies being blue, and the clouds being white. I remember kicking up my legs behind me as I ran. Everything felt and smelled fresh, and I was very happy.

-Age 4, East Montpelier, Vermont, USA

5

Emotional Memories

Overview

I hope you enjoyed that beautiful collection of memories! What caught your attention about those earliest childhood memories that were shared? Did you notice the strong emotions that were described? Well, that's no coincidence—emotions are one of the primary elements that leave a lasting impression when children (and adults) form memories! In this chapter, we highlight the emotional memories featured in the collection and then examine the role of emotions in memory processing.

First, we'll dive into the emotional themes found in the *Earliest Memories* collection, starting with unpleasant emotions then moving to pleasant emotions. I use the term *pleasant emotions* for feelings that are often called "positive," such as happiness or excitement, and *unpleasant emotions* for feelings that are often called "negative," such as sadness or anger. Emotions are functional, necessary, and each beautiful in their own way. They are not inherently positive or negative, good or bad, though they are often labeled that way. Now, let's explore the chills and thrills of the earliest memories from the collection.

Unpleasant Emotions in the *Earliest Memories* Collection

Several themes of unpleasant emotions arose from the collection, including fear, melancholy, self-consciousness, and frustration. Here, we'll explore these emotional experiences, as well as memories of crying, which can be associated with many different emotions.

Fearful memories ranged from mild unease to intense fright, including feelings of fear, intimidation, anxiety, and terror. For example, we read memories of feeling intimidated by worms crawling on the sidewalk, anxious about getting caught by one's mother, and terrified of a huge dog barking in one's face. These early encounters with fear shape our development of courage and coping strategies as we learn to navigate our anxieties.

Melancholy includes feelings of sadness, disappointment, and loneliness. For example, we read memories of feeling sad about not being able to eat a beaded cookie, disappointed about waiting in long lines at Walt Disney World, and feeling homesick and alone at the hospital. Sad memories involve loss, let-down, or longing and teach us that life is not always rosy. But there is a special beauty in the human experience of appreciating something we love, sometimes even more after it is lost.

Self-consciousness includes feelings of embarrassment, shyness, shame, guilt, remorse, and humiliation. Broadly, these emotions involve self-evaluation accompanied by timid or penitent behavior. We read memories of feeling embarrassed about trying to eat a wax candle, shy about wearing swimming shorts at the public pool, and guilty and regretful after cutting a toy pony's mane. These memories capture the emotional weight of being exposed or judged, leading to self-reflection.

Frustration includes feelings of irritation, annoyance, anger, and hate. For example, we read memories of feeling irritated at getting pushed in the sandbox, annoyed at being called home for dinner, angry at having a reward stolen on the school bus, and hating a rural area full of insects. These memories show the spectrum of displeasure we feel when we face perceived mistreatment or obstacles that thwart our desires.

Finally, there were several earliest memories of crying. Children cry for many reasons, such as expressing hunger, pain, discomfort, fatigue, sadness, fear, or frustration. Tears often say what words cannot. Crying communicates to others that we need care or attention and also helps us self-soothe and release stress.[57-59] We read memories about children crying because they got their head stuck between the staircase bars or their friend didn't say hello at school and weeping at hearing their mother's voice when feeling homesick. Tears are indicators of strong emotion, watermarks of poignant moments that stick in our memory.

Summary

We find many common childhood experiences in these unpleasant emotional memories. How many of us have felt that terrible anxiety about getting in trouble or that annoyance at having our playtime cut short? Childhood is a time of exploration, not only of the world around us but also of our internal world. It is a critical time when we begin to understand our emotions and place meaning on our experiences, largely based on how they make us feel. These foundational moments influence how we experience and express emotions later in life. Next, we'll shift into

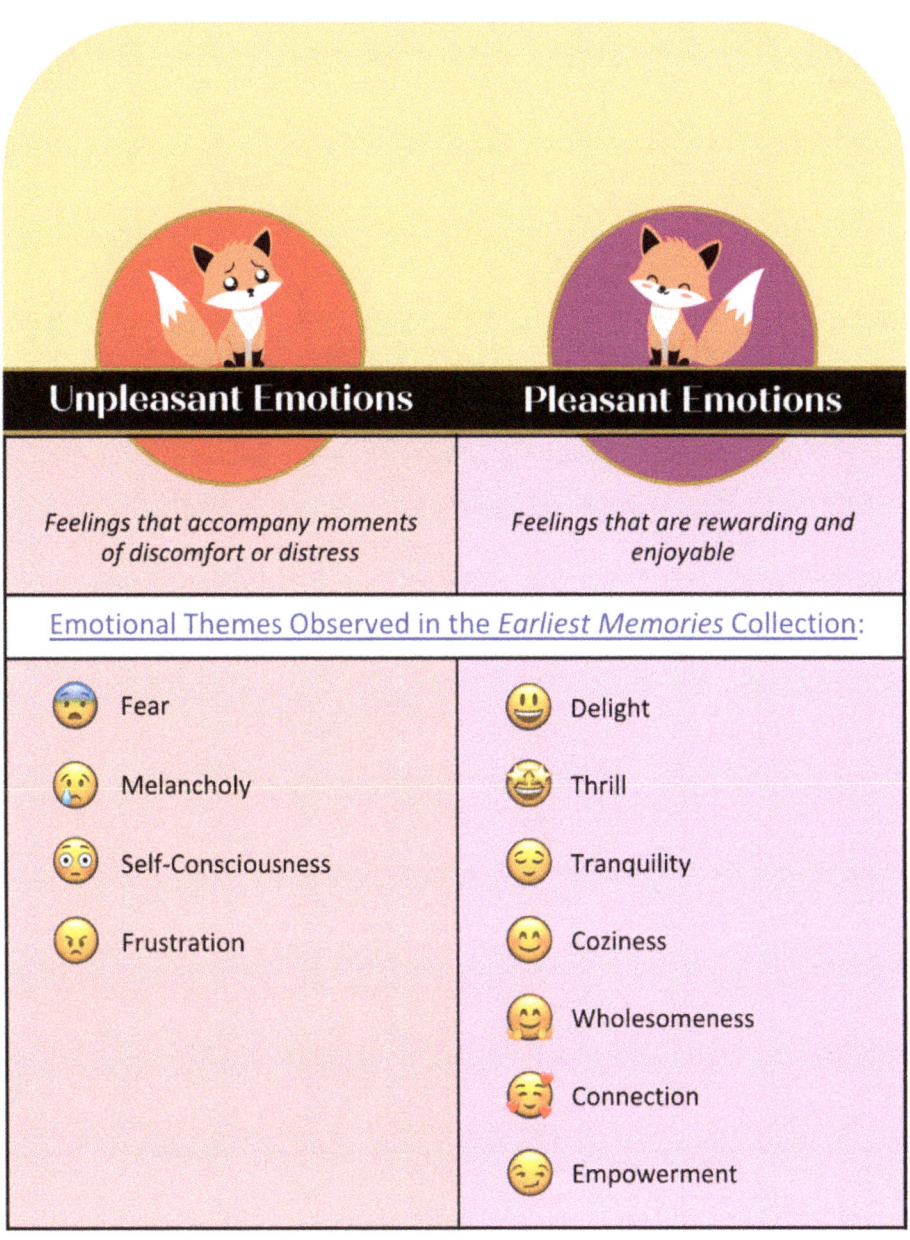

the realm of pleasant emotions and relive the joys of childhood.

Pleasant Emotions in the *Earliest Memories* Collection

In the collection, we read many pleasant emotional experiences, which can be grouped into common themes of delight, thrill, tranquility, coziness, wholesomeness, connection, and empowerment.

Delight captures feelings of happiness, fun, amusement, and enjoyment, along with memories of smiling and laughing. For example, we read memories of the happiness of catching a fish for the first time, giggling while being spun in circles, and having fun playing video games at the arcade with friends. We can feel the pleasure and sheer delight in the reliving of these uplifting childhood moments.

Thrill includes feelings of excitement, anticipation, curiosity, surprise, and exhilaration. Thrilling emotions captivate and energize us. For instance, we read about the surprise of getting a new bike on Christmas morning, the exhilaration of running in the house, and the excitement of being at Disney World. These memories all share a sense of heightened engagement, energy, and emotional intensity. When we are young, we can find excitement in the simplest things, such as going to a cafe and sitting on red leather seats. The thrill of novelty creates exciting moments that make us feel vibrantly alive.

In contrast, tranquility is a gentle, serene emotion. This theme includes feelings of safety, peace, and calm. For example, we read memories of children feeling safe and secure after being rescued and feeling peaceful rocking in their mother's lap. Tranquil experiences center and ground us as we find peace in inner stillness. For children, the feeling of safety they experience when living in tranquility is crucial for their emotional growth. It provides a calm baseline for the development of a regulated nervous system.

Feelings of coziness include the warmth, ease, and security that unite physical and emotional experiences of comfort. For example, we read memories of feeling warm and cozy in a crib with a teddy bear and blanket, being given a warm bath by one's aunt, and being snuggled up against one's mother. These memories paint a picture of the profound contentment that comes from feeling safe and snug. For children, the feeling of security that coziness brings is rooted in the care provided by family members. The presence of attentive caregivers serves as an emotional anchor, laying the groundwork for a secure emotional base.

Wholesomeness includes feelings of innocence and pureness that come from being in harmony with oneself and the world. We read about the pure and authentic feeling of sitting on a parent's lap to read a book or watch the morning news, the innocence of sitting on the ground playing with toy ponies, and the heartwarming simplicity of making homemade play dough at grandma's house. These wholesome memories capture the goodness and sincerity that are present in us all from childhood.

Connection includes feelings of affection, love, unity, and camaraderie. Connection is a deep and meaningful sense of closeness that bonds us with others through our shared human experience. For example, we read memories of a child feeling loved by her doting grandparents, the special feeling of a romantic gaze, and the heartfelt joy of cuddling with a new puppy. These memories depict the warm sense of belonging that comes from being closely connected to others in a meaningful and positive way.

Empowerment includes feeling power, freedom, determination, and a sense of control. Empowerment is the uplifting emotion that comes from feeling capable, strong, and independent. We read memories of children feeling the freedom of sneaking out for a neighborhood adventure, feeling determined to disassemble a toy plane, and feeling powerful wearing a poofy jacket atop the jungle gym. These stories retell experiences of strength and autonomy as children exert influence over their environment. During childhood, we often have limited control over the world around us. Thus, the ability to explore, express, and assert oneself feels extremely liberating, and plays a vital role in shaping a child's sense of self-efficacy.

Summary

These intimate memories from the collection show the breadth of emotion we experience during childhood. Whether it's the sweet warmth of a new crush's smile or the utter heartbreak of losing our prized crayons down the commode, our emotional memories naturally feel significant to us. These moments help us appreciate the richness of our emotional lives and our incredible ability to feel our way through life's ups and downs.

The Role of Emotions

Our emotions serve an important function—they trigger us to respond to our environment and take actions that will ensure our survival. For example, fear prompts you to leave a dangerous environment. The memory of that fear is also protective as it may prevent you from returning to a dangerous place in the future.

Our body gives several indicators to signal that we're having an emotional response. When something happens to us, we feel physical sensations in our body, often accompanied by an emotional experience. For example, if we lose a friend, we may experience a tightness around our heart, a heaviness in our limbs, and a feeling of sadness. If we are mistreated, we may experience heat rushing to our head, tension in our muscles, and a feeling of anger.

Emotions can also stem from our interpretations of a situation. We feel sad when we *perceive* that someone is pulling away emotionally or angry when we *assume* that someone had malicious intent. In this way, perception is reality. Our body's response aligns with what we believe.

Emotions and Social Bonds

Emotional memories are flagged as "important," especially when they are necessary for our survival. After all, memory and emotions both evolved to keep us alive. In prehistoric times, the world was a very dangerous place and we relied on our tribe to protect us. Social rejection would have been *deadly!* So, if you did something to warrant rejection, feeling shame was a vital emotional response. Think about it… what do we do when we feel ashamed? Our natural instinct is to hide. We withdraw to engage in critical self-reflection. When we re-emerge, we may bow our heads in shame and try to make amends with those we offended.[60] These actions are functional ways to avoid punishment and rejection so we can preserve key social bonds and promote our own survival.

Even in our modern world, our emotions continue to serve as important messengers that move us to take action and protect ourselves. Remember the memory of a child being laughed at and turned away from a birthday party? They described feeling painfully embarrassed and ashamed for not knowing the party was just for family. While it was an honest misunderstanding, their shame was a natural reaction to the social rejection they encountered. Although painful—and some might say cruel—this interaction was a valuable opportunity for social learning. Those strong emotions catch our attention, prompting us to develop adaptive strategies to navigate complex social situations in the future.

Emotions and Memory Processing

Because our emotions are so vital for our survival, our brains process emotional memories differently from other memories. Our memory for emotional events is enhanced during all three stages of memory processing: encoding, storage, and retrieval.[61] Think about it—a hilarious movie that makes you laugh and a heartfelt movie that makes you cry are often more memorable than a neutral documentary. Let's break down why that is.

Imagine you're watching a thrilling movie that really tugs at your emotions! In your excitement, a chemical called norepinephrine is released, triggering your brain and body to react. This chemical sets off a series of reactions that helps you encode what's happening in the moment and also strengthens your memory storage over time. In this way, there is a coordinated effort between the parts of your brain that handle your emotions and your memories, which helps you form a stronger memory of the event.[61-64] Meanwhile, your emotions keep you glued to the movie, paying extra close attention so you don't miss a thing! This increased focus during emotional moments enhances encoding and yields more vivid memories.[65,66] Overall, the heightened emotions, the focused attention, and the collaboration between different parts of your brain all result in stronger memories for emotional experiences.[61]

Later, during memory retrieval, the parts of your brain dealing with memories and emotions team up once again. This coordination makes it easier for you to access the memory of the exciting movie and retain it over time. Emotions also serve as important cues that help prompt our memory.[61,67,68] That buzzing excitement you feel when telling your friend about the film later mirrors how you felt while watching it. This emotional match makes it easier to remember the movie.

Do you remember the memory experiments in Chapter 2 with the deep-sea divers? We learned how context provides cues to help us retrieve stored memories.[15] Well, context is more than just your physical environment—it also includes your social setting and your mental and emotional state.[69] Our current emotional state impacts what we remember—so, when we're in a good mood, we tend to think of positive memories, and when we're in a rotten mood, we're more likely to think of negative memories.[70] Overall, our emotions have a strong impact on how our memories are formed and how we recall them later.

Emotional Quality and Intensity

Now that we've established the importance of emotions in memory processing, we'll consider how different aspects of an emotional experience influence what we remember. Let's discuss how emotional intensity and tone can shape our autobiographical memory.

Strong emotions increase our memory—not only for the emotional parts of the memory, but also for other neutral details.[61] For example, think of something major that you've accomplished, such as graduating from high school. For most people, their high school graduation was a significant emotional experience that is not easily forgotten. Thinking back on that day, you probably remember feeling mixed emotions as you walked across the stage, such as pride, excitement, and nervousness. You may also remember other details from the day, like the weather or what you wore. Those details may stick in your mind more than the weather or your attire on any given day because they are tied to a significant emotional event.

Aside from instances of extreme stress, the more emotionally-charged an event, the more likely it is to stick in our memory! In fact, the strength of our memory depends more on the *intensity* of the emotion than on any specific *type* of emotion.[62,66] So, when it comes to memory, feeling completely heartbroken versus slightly somber matters more than whether we're sad or angry. Intensity takes the cake!

Nonetheless, specific emotions *do* impact memory in distinct ways because we instinctively attune to emotionally relevant aspects of a situation. For example, people with social anxiety anticipate being judged negatively by others, so they tend to hyper-focus on the negative details of a social interaction. This selective attention leads to selective memory, so they may remember a date's furrowed brow but not their welcoming smile. Each emotion makes us hone in on a different type of information, so you'll remember an experience differently depending on how you were feeling at the time.[71]

Fear motivates us to escape danger, so when we're afraid, we strongly encode information that will help us avoid future threats. Anger motivates us to overcome obstacles, so when we're angry, we capture information that can help us clear barriers to our goal. We feel sad when we experience a loss that we believe we cannot change. Thus, sad memories often focus on the consequences of what happened, so that we can readjust our lives and adapt accordingly. When we think back on a sad memory, it's natural to wonder whether there was anything we could have done differently to change the outcome. That is our brain trying to prevent future loss. It doesn't mean we did anything wrong—our minds simply process loss by striving to improve our chances going forward.[71]

We also tend to remember happy events better than neutral ones.[72] During happy moments, the pleasure you experience boosts memory encoding and storage. It's like our mind's way of saying, "This is good— let's remember this!" When you reminisce about a positive memory, the parts of your brain that process emotion and reward are activated and you essentially re-experience the pleasurable sensations and feelings associated with that event.[73] For example, close your eyes and think about the most delicious dessert you've ever eaten, whether that's a gooey chocolate lava cake or a creamy panna cotta. Imagine slowly licking it off the spoon and savoring its sweet taste and silky texture. Notice the pleasure you can generate simply by re-engaging with a memory.

Research has shown that over time, emotions associated with unpleasant memories fade faster than pleasant memories.[74] Thus, for most people, happy memories stick around longer. This is good news, as a tendency to remember things positively is associated with greater life satisfaction![75] Positive and negative memories also vary by age. A group of researchers combined results from over 100 scientific studies and found that older adults remembered more pleasant than unpleasant events, whereas younger adults' memories were more negative.[76]

So, what do we remember better overall: positive or negative experiences? This common question has a nuanced answer. Positive experiences tend to be remembered more quickly and with greater contextual detail, whereas negative experiences are remembered more vividly and with greater detail for the core event.[68,77] For example, you may have a rosy, big-picture memory of a typical summer day at the lake. But if you lost your diamond ring while swimming, that dramatic detail would dominate your memory of the day.

The Impact of Stress on Memory Processing

Stress impacts our memory in several different ways. During high-stress situations, our body goes into "emergency mode," releasing cortisol and other stress hormones. These hormones act on our memory processing system and have different effects at various stages of memory processing.

Stress impairs our memory during retrieval—you may have had this experience if you've ever "drawn a blank" on a high-stakes exam. Interestingly, stress has the opposite effect on memory formation. Stress enhances encoding and storage of the core emotional aspects of an event, yet this often comes at the cost of remembering the peripheral details later.[77,78] Thus, it is common for people to remember the key components of a crisis, while feeling like some of the finer details are hazy in their memory. For example, when someone is robbed at gunpoint, they focus on the most prominent aspect (the gun) and may not clearly remember what the robber looked like. As you can imagine, this has serious implications in forensic settings, especially when a case relies on eyewitness testimony.

As Harvard neuroscience professor, Elizabeth Phelps, explained it (Steig, 2019[79]):

> The memory wasn't designed for the legal system. Memories were not designed so that we can accurately recollect the past and all of its details. Memories were designed to allow us to do adaptive things in the future, to tell us the right things to do.

Forensic Case Example

I served as a jury member on a very challenging case involving the alleged assault of a young girl by her family friend. In this case, there was little evidence presented aside from the testimonies. However, by state law, an assault victim's testimony alone was sufficient to support a conviction. So, there we were, twelve strangers in a room trying to come to a consensus on whether this young girl's story was credible. Determining the validity of her testimony was a huge responsibility for the jury, and reaching a verdict was an excruciating and emotionally exhausting experience for everyone involved.

We heard several accounts of the story: from the alleged victim herself, her mother, the police report, and the recording of an interview conducted by a forensic child psychologist. The alleged perpetrator (her family friend) did not testify. There were some inconsistencies across the different accounts. Were they in the living room or the kitchen when they began the conversation leading up to the assault? Eventually, they end up in the kitchen. In the original interview, the girl said she got a soda out of the refrigerator, but she did not mention that on the witness stand three years later…

I am not trained in forensics, but as a psychologist who has conducted countless hours of trauma-focused therapy and assessment, I found these small changes and omissions to be completely normal. As discussed earlier, people often do not pay attention to peripheral parts of a highly emotional event, so minor details may not endure in their memory. The hearing took place several years after the initial incident and it is common for small details to be lost over time or modified during memory reconstruction. Further, the way a question is posed—by a forensic child psychologist versus an attorney, for example—can impact how a child tells their story.

My professional perspective on memory was met with skepticism from other jury members. *"She's lying!"* some said of the young girl. *"She keeps changing her story."* After the second day of deliberation, many jurors had deemed her an unreliable witness. And despite our earnest efforts, we could not come to a unanimous decision. The case was declared a mistrial and would be retried with a new jury.

Somatic Memory

Memories are stored not only in the brain, but also as physical responses held deep within our bodies. This is called *somatic memory*. Our bodies hold somatic memories from major traumatic events, as well as everyday stressful situations.[80-82] For example, getting cut off in traffic or hearing someone shout are experienced physiologically as threats by our stress response system.

Many of us do not allow ourselves to fully feel our emotions and release them through our body's natural responses, such as crying or shaking after a stressful experience. Instead, many people suppress their body's responses, such as distracting themselves with work and entertainment or analyzing the experience logically instead of feeling their emotions directly. When we do not process stress through to a full physiological and emotional release, that stress is stored in our system and held physically in our body's tissues. Our nervous system can become tuned to this heightened stress level and may fail to return to a calm baseline. Thus, we remain in a state of high alert and, eventually, our body goes into a physiological shutdown to cope with the constant stress.[81]

Living in a state of chronic stress can lead to a constellation of physical, emotional, and mental health issues. Fortunately, therapeutic techniques that focus on physiological processing in the body, such as somatic therapy, can help restore balance to the nervous system and promote healing. Somatic approaches provide a framework for healing that helps us process and integrate our memories, not just mentally, but throughout the entire body.[81]

We begin to heal when we can stay fully connected to our body while re-engaging with stressful memories, allowing the body's natural response to arise. Our body can then discharge stored stress. Each person releases stress in a different way. It may feel like tingling sensations, temperature changes, body movements, muscle cramps, vocalizations, or feeling a strong emotion. When we release accumulated stress, this reduces the burden on our nervous system and frees up our resources to respond more flexibly to life's ups and downs.[81,82]

The body and mind mirror one another—as we reconnect to our bodily responses, our mind is also able to make sense of past experiences. We form clear, organized narratives about our past and gain insight into our role as one piece of a broader context. We can accept what we have been through and integrate our memories into a deeper understanding of ourselves and the world around us.[83,84]

Closing Thoughts

Emotional memories connect us to the collective human experience. While the exact details may vary, the emotions we feel remind us that, despite our differences, we hold common threads of joy, sorrow, love, and hope. Our emotional memories enrich our personal lives and serve as a bridge to the experiences we share with others.

In the next chapter, we turn to sensory memories, which were also prominent in the *Earliest Memories* collection. As you will see in the memories we highlight, sensations and emotions are tightly linked. As humans, it is our nature to react to what we experience with our senses: Sensation leads to perception, perception to evaluation, and evaluation to emotion. Therefore, as you learn about sensory memory, remember that the emotional significance of these sensations also impacts memory processing. Stronger emotions create more vivid, long-lasting memories.

6

Sensory Memories

Overview

Childhood memories are often filled with rich sensory details, as showcased in the vibrant experiences of sight, sound, smell, taste, and touch captured in the *Earliest Memories* collection. In this chapter, we take a journey through the five senses to explore why sensory details stand out so vividly in our long-term memory.

Sensory Memories

You may recall from Chapter 3 that memories from before the age of five are typically recalled as "fragments" because younger children have not yet developed the language abilities and other cognitive structures necessary to create more complex memories.[22,37] Early memories commonly lack the narrative structure of language, proceed without a clear order of events, and typically look like "snapshots" or simple moments in time.[85] These unformed memories in early childhood therefore reflect the immature cognitive and language capacities of young children during this developmental period.

In early childhood, we are first and foremost sensory beings. Before we develop more complex abilities such as abstract thinking, we are more reliant on direct experience through our senses to understand the world around us. In this way, very young children live more fully in the present moment. They are less prone to mental time travel than adults who tend to worry about the future or dwell on past mistakes. Young children are *right there* in their bodies, taking in their current environment with their senses, paying attention to what they see, hear, taste, touch, and smell to make sense of their daily lives.

Sensory memories create a natural type of intimacy that is rooted in presence. Hearing someone describe the precise sensations they experienced transports us into their universe. It drops us down into that moment with the person as we explore the world alongside them. The excitement and innocence of childhood memories evoke that nostalgic feeling of seeing the world with new eyes. In the *Earliest Memories* collection, we get to experience the stretchiness of pizza dough for the very first time, the satisfying sensation of brushing your fingers along a roughly textured wall, and the first whiff of chlorine from a day at the pool. We are given the chance to relive these precious human moments through the simple curiosity of a young child.

Now, we'll explore the different sensations described in the collection through sight, sound, smell, taste, and touch. First, let's dive into our sense of sight!

 Sight

Visual Memories

When we are young, the visual world around us is *astounding!* There are so many novel things to look at, so many interesting details to drink in with our eyes. Vision is a large part of what makes our memories so vivid. This becomes even more apparent when we examine the visual elements described in the *Earliest Memories* collection.

Color

It's remarkable how *precisely* we can remember specific colors years later. In the collection, we read memories of the blue Smarties dye streaked through cake cream, the orange packet of Sainsbury's crisps, and the green plaid cot for a trip home from the hospital. As humans, we are wired to notice color, which played a crucial role in our survival. We relied on color to identify ripe fruits and avoid dangerous snakes. Color is distinctive, making objects stand out from their surroundings and capturing our attention, enhancing our memory.

Visual Movement

Many memories in the collection featured captivating visual movement: trees swaying in the wind on a pram ride, a plastic crescent moon peeling off the ceiling, worms crawling on wet pavement, and wind chimes swaying on great grandma's back porch. Like color, movement quickly captures our attention. Motion stands out against a static background—it signals that something is happening and we should pay attention!

Sunlight

Sunlight was another common element in the collection: a sharp line of sunlight across the wall, a glimpse of sunlight through transparent curtains, and sparkling sunlight dancing on

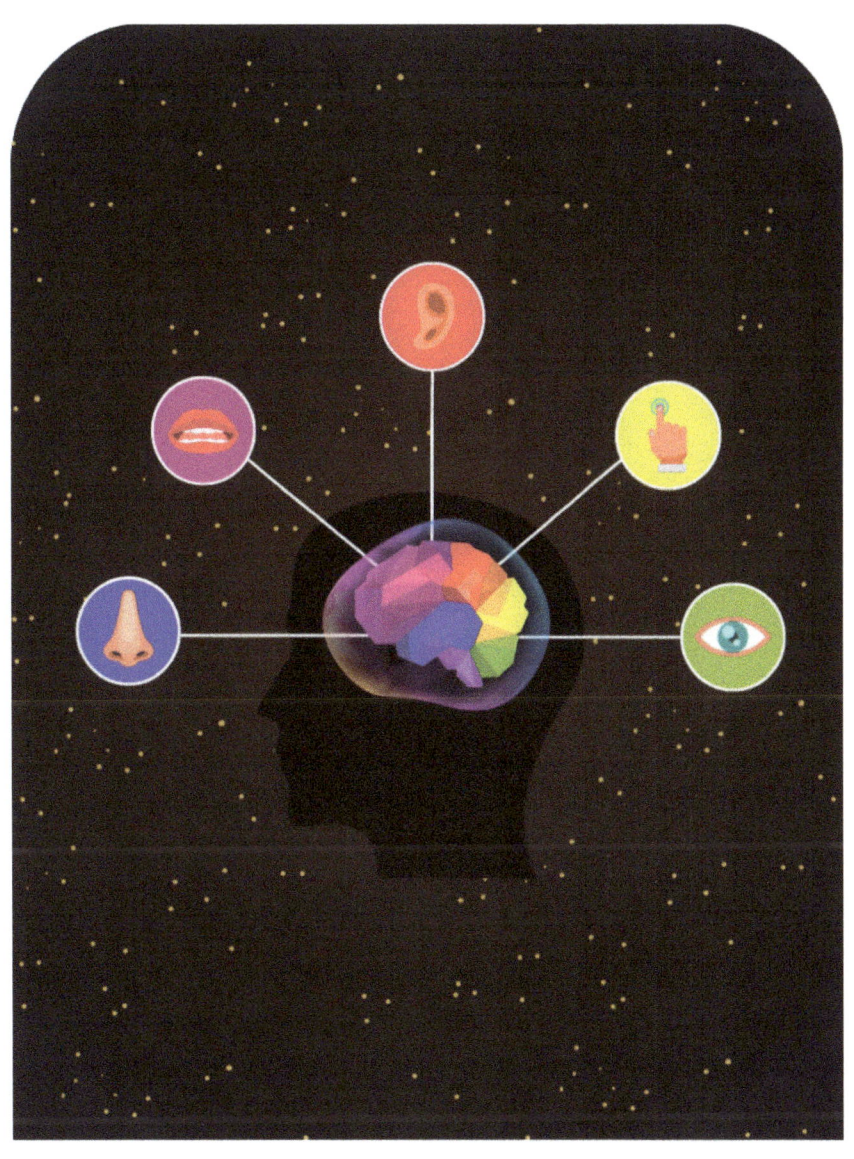

the waves in a pool. Sensations of warmth and light are significant in our early memories. Sunlight nourishes us, awakens our vital life force, and reminds us of the radiance of life. Light has the power to transform an otherwise mundane occurrence into an absolutely breathtaking experience that makes a long-lasting impression.

Visual Reflections

We read several memories of children noticing their own reflections. When we are young, we spend most of our time looking at other people and the world around us. Thus, the experience of seeing ourselves can be quite striking, especially under meaningful circumstances! Take, for instance, the memory of a child seeing herself snuggled up safely against her mother in the reflection of the bathroom mirror. Or a young boy looking over a deep trench and seeing his own petrified face in the reflection of the water below. These meaningful moments make us suddenly aware of ourselves: our image, our relation to others, and our fragile place in this world. The self-awareness we gain through exposure to our own facial expressions amplifies our emotional experience, which can imprint deeply on our memory.

By the time children reach 18-24 months, they can typically recognize themselves in a mirror and begin to develop an early self-concept. When faced with their own image, toddlers are already starting to view themselves through the eyes of others. Over the next few years, they develop an awareness of self that continues to build over time.[86] This sense of self facilitates children's autobiographical memory. Both children and adults remember things better when they are relevant to our self.

Memory is stronger for personally meaningful information, in part, because it is easier to process and organize during encoding. It's as if the brain puts a "self" label on the memory and then knows exactly where to file it—in the folder of experiences related to me.[87,88]

Emotion Identification

Finally, as children develop emotional awareness, they also learn to use visual cues to identify how others are feeling in a moment's time. Like the "scary look" on a mother's face while rescuing her child from a near drowning. Or the positive energy emanating from a sister's smile while chatting with her younger sibling. In this way, body language and facial expressions capture emotional states in our visual memory. As we will continue to explore in this chapter, our sensations are intimately tied to our emotions.

Sound

Auditory Memories

Sounds play a significant role in shaping our memories. We heard a broad range of sounds in the memories that were shared. Some sounds were distinctive and jarring, such as the barking of a neighborhood dog, a loud crack of thunder, or a spoon rattling across the floor during an earthquake. Some sounds were playful and pleasant, such as the chirping of the birds in a forest. Some sounds were soothing in tone, such as the soft sound of a mother's voice, or in rhythm, such as the comforting noise of a fan blowing. The simple consistency of a rhythmic sound has a predictability that calms us, akin to the repetition of a lullaby or the sensation of being rocked back and forth, back and forth.

Background Noise

Background noises set the ambiance and context of our memories, like the exuberant sounds of a Mardi Gras parade in New Orleans or the sounds of chains pulling and wind blowing on a rollercoaster ride. When we are relaxed and taking in the environment around us, we notice these background noises more readily. But when we need to focus, our mind naturally filters out noises that are not relevant to the task at hand so that we do not get distracted.[89]

Compared to adults, children tend to listen more broadly to background noise and have less selective focus on a single source, such as someone speaking to them.[90,91] Parents, no doubt, will be unsurprised by this fact! Although the basic human auditory system is well developed by about six months of age, children's cognitive abilities are not fully developed. Children cannot differentiate simultaneous sounds as well as adults, making it more challenging for them to focus on one sound.[91] Researchers typically describe this as children's "failure" to tune out distractions. But in a world that struggles to stay fully present, the ability to be mindful of the surrounding environment is one of the beautiful gifts of childhood. Each stage of life brings its unique trials and triumphs.

Emotion Identification

Hearing can help us understand and encode emotions in early

memories. From infancy, humans are sensitive to vocal cues of "positive" and "negative" emotions, such as happiness versus sadness.92 Throughout childhood and adolescence, we become more and more skilled at recognizing emotional nuances expressed through sound.93 Further, emotions enhance our auditory memory: by school age, children already remember emotional sounds better than neutral sounds.94

Consider, for example, memories of the soft sound of a mother's voice on the phone or the concerned tone of family members' rapid inhales ("ahhhhh"). Children are very attuned to this type of vocal communication that runs deeper than words. They naturally look to their caregivers to set the tone and most kids learn quickly what is valued by adults—and what is not.

 Smell

Scent Memories

Scents have the power to evoke extremely vivid memories. In the *Earliest Memories* collection, we read descriptions of various childhood smells, like the aroma of flowers at the park, the smell of homemade bread baking at grandma's house, and the scent of leaves and grass in the forest on a school trip. These olfactory traces are woven into the network of associations that make up these early memories. With such rich scent memories, it is unsurprising that smell is the human sense most closely linked to memory! This is known as the *Proust phenomenon*, stemming from a famous passage by French novelist, Marcel Proust. In this writing, Proust describes his experience of drinking a spoonful of tea after soaking a petite madeleine cake in it (Proust, 1960/1922₉₅):

> No sooner had the warm liquid mixed with the crumbs touched my palate than a shudder ran through me and I stopped, intent upon the extraordinary thing that was happening to me. An exquisite pleasure had invaded my senses, but individual, detached, with no suggestion of its origin.

This fleeting sensation transports Proust to a corner of his mind that he cannot place and creates an insatiable drive to pin down that old experience. Over and over again, he tries to drag the memory up from the "abyss" but is unsuccessful. When he finally gives up and allows his mind to wander as he drinks his tea, it comes back to him in an instant: (Proust, 1960/1922[95])

> [...] And suddenly the memory revealed itself. The taste was that of the little piece of madeleine which on Sunday mornings at Combray (because on those mornings I did not go out before Mass), when I went to say good morning to her in her bedroom, my aunt Léonie used to give me, dipping it first in her own cup of tea or tisane.

Suddenly, Proust's mind was flooded with images of his aunt's grey house where he used to eat tea-soaked madeleines as a child. These images expanded to the village, the surrounding gardens, and the entire era of childhood where the memory took place.

The *Proust phenomenon* is real, though nuanced. Smells do, in fact, provide the strongest cues for our memory of personal experiences. Compared to other senses, olfactory memories are more vivid, more emotional, and reach back earlier in childhood, but they also tend to be rarer.[96,97] The olfactory center is part of the limbic system, a set of structures in our brain that plays a major role in processing memory, emotions, and... scents! Compared with our other senses, the olfactory system has the fastest and most direct connection to our emotions. Thus, smells evoke quick emotional responses and summon strong associations.[98]

Revisiting A Proustian Experience

When I was in high school, my family took an epic road trip and drove from Louisiana all the way to the Grand Canyon in an RV, stopping at campsites and national parks along the route—it was pure Americana! From visiting the White Sands National Park, which looks like something out of an alien planet in *Star Wars*, to rafting along the Colorado River at the base of the Grand Canyon, this was one of our most memorable family trips.

Throughout most of my adolescence, I had used Burt's Bees lip balm, but for this particular vacation, I got a tropical lip balm called Panama Jack. I was constantly reapplying it to protect my lips from the arid desert air. That smell became absolutely *gridlocked* in my memory in association with this Grand Canyon trip. When we wrapped up our adventures and returned home to everyday life, I resumed my loyalty to Burt's Bees. But for years, I kept that tube of Panama Jack that had the tiniest trace of lip balm left at the bottom. From time to time, I would pull it out, breathe in that specific smell, and find myself blasted back to the grand beauty of the Southwestern United States, engaging in a ritual with that intangible experience of olfactory memory that Proust had stumbled upon many years before.

What is a nostalgic smell from your childhood? Whether it's

the powdery, floral fragrance of your mother's perfume or the oddly comforting smell of the plastic mats you laid on during naptime in kindergarten, take a minute to close your eyes, inhale deeply, and see if you can revisit the smells and sensations of those early experiences.

Taste

Gustatory Memories

Next, we turn to gustatory memories, which relate to our sense of taste. If Proust's cake-crumb tea cued his childhood memories, then why does the *Proust phenomenon* refer to smell instead of taste?

The distinction lies in *taste* versus *flavor*. There are five tastes that are registered by our taste buds: sweet, salty, sour, bitter, and umami. Flavors, on the other hand, are the specific notes that we recognize in a food, such as raspberry or licorice flavor. On its own, our gustatory system can only recognize the five basic tastes. It works together with our olfactory, memory, and emotion-processing systems for us to identify and experience different flavors.[99] That's why Proust's famous madeleine memory was largely based on smell—he did not simply taste the sweetness of the cake, he also remembered the complex flavors of cake soaked with lime-flower tea that stemmed from his sense of smell.

Sweet Memories

We read a couple of earliest memories related to taste, such as enjoying the sweetness of a bottle of warm chamomile tea, or the sweet pumpkin pie and rich chocolate frozen yogurt at

Thanksgiving. While taste-based memories are uncommon,[100] those that do persist from childhood typically involve sweet foods because they activate our brain's reward systems.[101] These reward systems, in turn, are implicated in long-term memory formation and can facilitate encoding, as well as the integration of memories into our existing knowledge during storage.[102,103] Sugary foods are naturally rewarding and are often associated with pleasant emotions. Especially for children, sweets tend to be used for rewards—like picking out a candy bar at the grocery store checkout for being well-behaved—or during celebrations, such as enjoying chocolate gelt during Hanukkah.

 ## Touch

Haptic Memories

Lastly, we turn to tactile, or *haptic* memories, which involve our sense of touch.[104,105] There were a wide variety of tactile memories in the collection, such as remembering the cozy texture of a blanket, the texture of the gravel in a turtle's bowl, or feeling one's legs against the wooden slats of a bench. There were also tactile memories of temperature like the steamy greenhouse on a walk with Nanna, the coolness of the pool on a hot day, or the cold wind on one's face while sledding in the driveway.

These haptic memories show children's sensitivity to different textures, surfaces, and temperatures in their surroundings. These examples again underscore how important basic sensory experiences are to young children as they learn to navigate their world. Touch allows us to explore our surroundings, to give and receive social connection and communication.[100] The skin is the largest organ of the human body and thus, our primary connection with our external environment. Our sense of touch puts us in direct contact with the world around us.

Kinesthetic Memories

Haptic memories also include kinesthetic sensations, or the way we move and experience our bodies in space.[105] We read numerous kinesthetic memories, such as kicking one's legs up while running across the yard, kicking a soccer ball, and peeling bits of detergent off a bottle. In childhood, we begin to build the repertoire of movements that we will use throughout our life. Some motions, like walking, reaching, and pulling are repeated so many times that they become engrained in our muscle memory.[106]

As children develop, their movements become more complex. Over time, coordinated sets of movements become locked in our memory, such as writing your signature. As a well-practiced adult, once you begin signing a paper, your muscles fall into the flow of the movement that you know so well—it is automatic. This is not a memory you access with your mind voluntarily. Rather, your body knows how it physically *feels* to write your name or brush your teeth.[106] Our life story is not only written in words or pictured in images—it becomes etched into our body by the movements of our daily life.

Pain and the Mind-Body Connection

Finally, we read several examples of painful childhood memories, like the sharp jolt of a friend's head striking one's nose or the throbbing ache of a bee sting. Painful memories are particularly potent, and it can be difficult to disentangle the physical versus emotional components of pain. For humans, pain represents a complex overlap between a physical sensation and an emotional experience. Psychological and bodily pain are connected in many ways. To name just a few:

1) There is substantial overlap in the network of brain regions that process physical and emotional pain.[107]

2) Physical and emotional pain influence one another directly. For example, holding the hand of a loved one can reduce feelings of physical pain.[107]

3) Physical and emotional pain produce a similar reaction in our body. For example, they both induce our body's natural pain-reducing reaction.[107]

4) Physical and emotional pain respond to similar treatments. For example, over-the-counter painkillers, such as acetaminophen, can reduce social pain like the pain of feeling rejected.[108]

Common metaphors like "feeling torn" or "experiencing heartache" are not just poetic; the stab of betrayal and the sting of rejection are real! When a difficult social situation hurts us, it really *hurts* us because the experiences of emotional and physical pain are processed similarly by the brain and our memories often reflect a high degree of overlap between the two.

Painful Memories

Pain stands out in our memories. Studies have shown that we are able to recall the intensity of past pain with reasonable accuracy.[109] Think of a common painful experience from childhood like scraped palms from falling on rough concrete. Can you recall that stinging sensation? Can you feel the heat and swelling as blood rushes to your hands?

The way children recall painful events and discuss them with their parents can impact their future responses to pain. Positive framing can promote healthy coping and reduce sensitivity to future experiences of pain, such as medical procedures.[110] For example, adults' direct discussion of pain with children, asking open-ended questions, putting pain in context, and elaborating with emotional language, can help kids learn to regulate their own emotions around pain, as well as increase their empathy for others' pain.[110,111]

Closing Thoughts

Sensory experiences play an important role in enriching our memories. Early in life, sensory exploration serves as a primary avenue for learning and development, laying the foundation for more complex cognitive processes. Memories associated with vivid sights, sounds, smells, tastes, and touches are easily accessible and often evoke deep emotional connections. These multifaceted experiences are highly engaging and thus, more memorable: sensory-rich memories form lasting impressions from childhood and beyond. Sensory memories also feel nostalgic because revisiting those specific sensory details allows us to reexperience the moment, connecting us more closely with our past. In the closing chapter, we explore the bittersweet beauty of nostalgia and why our earliest memories tug at our heartstrings.

7

The Significance of Our Earliest Memories

Overview

This final chapter covers the content themes from the *Earliest Memories* collection and explores why certain memories stay close to our hearts. We'll dive into the universal experiences present in these first childhood memories, what makes our earliest memories so special, and the nuanced nature of nostalgia.

Earliest Memories Collection Themes

A number of universal themes are found in the *Earliest Memories* collection, including Family Connections; Nature and Exploration; Social Interactions, Play, and Learning, and finally, Adaptation and Transformation. Let's explore the psychological foundations of these formative childhood experiences.

Family Connections

The theme of Family Connections encompasses memories of interactions with family members and loved ones, including physical affection and emotional support. Many of us fondly remember childhood memories of family trips and activities, holidays, birthdays, and companionship with our beloved pets. For example, we read about first memories of a secret handshake shared between twins, sitting on a mother's lap in a rocking chair, comforting words from an uncle when feeling ill, and playing at the lake with one's cousin. These memories beautifully capture precious moments of joy, connection, and shared traditions.

Memories of positive interactions with others, especially with family, are more strongly related to a sense of happiness during childhood, compared with pleasant solitary activities like independent play.[112] Early family connections are crucial for the foundation of our emotional health and future well-being. *Attachment theory*, developed by John Bowlby and Mary Ainsworth, suggests that the bonds formed in early childhood significantly influence our psychological and emotional development throughout our lives. Secure attachment involves a strong emotional bond between a child and a caregiver who provides them with safety, security, and consistency.[113,114]

A secure attachment relationship with a primary caregiver during childhood serves as a cornerstone for well-being across many domains. Research indicates that secure attachment promotes emotional well-being, physical health, self-esteem, effective regulation of the nervous system and emotions, and positive social and cognitive development.[113-118] Further, our primary attachment forms the blueprint for our future relationships, shaping our expectations, communication style, and the way we approach connections with others later in life.[114]

The theme of Family Connections also includes memories of home and familiar environments, which provide comfort and a sense of belonging. For example, we read memories of waking up from a nap and feeling warm and content at home or visiting grandmother's magical yellow house amid the spice plantations. *Place attachment* refers to the emotional bonds we form with specific places,[119] such as our childhood room, favorite ice cream shop, or secret tree fort. Feeling connected to a familiar physical place helps children feel secure and grounded. It creates continuity in the part of our identity that is tied to that particular place.

For example, someone who lived in the same neighborhood throughout their formative years might feel a deep bond with the neighborhood park that served as both the playground for their youthful imagination and the sanctuary for contemplative walks during adolescence. Places connect us back with ourselves and with others—it's hard to imagine the halls of your elementary school without thinking of your old classmates or favorite teacher. Places provide a tangible context to contain the meaningful shared experiences that have shaped us throughout our lives.

Nature and Exploration

The theme of Nature and Exploration includes memories of outdoor exploration and engagement with nature, which fosters a sense of curiosity, adventure, and connection with the natural world. For example, we read memories of feeding ducks on a family vacation, knocking down mushrooms from a tree branch, and laughing at a squirrel on a school field trip.

Time spent in nature can promote children's development as they use their curiosity to explore. Children learn from the world around them through sensory-rich experiences, such as listening to bird calls on a hike or skipping stones across a pond. Hands-on exploration encourages physical activity and the development of motor skills. This natural education gives children a sense of autonomy as they make their own discoveries.[120] It also cultivates environmental awareness by providing opportunities for learning to respect other creatures and preserve the harmony of nature.[121]

Social Interactions, Play, and Learning

The theme of Social Interactions, Play, and Learning includes memories of playing with toys, engaging in imaginative activities, and interacting with peers. For example, we read stories of pretending to be one of the X-Men, playing with friends at Chuck E. Cheese for a birthday party, and playing hide and seek at the playground with neighborhood friends.

Psychologists who study early child development emphasize the importance of play and its profound impact on cognitive and social development for young children. Play provides so much more than mere amusement. As children immerse themselves in imaginative play through self-directed activities, interactions with peers, or guided play with a knowledgeable adult, they are building key developmental skills.[120,122]

For example, children often create imaginary scenarios when playing, such as pretending to be a friend's mother while chatting on a pinecone phone. This type of play helps children develop abstract thinking skills, practice creativity and cognitive flexibility, engage with social rules and cultural norms, understand others' perspectives, and participate in social problem-solving and negotiation.[120,122,123] Play sets the stage for children to "learn by doing" through imaginative improvisation.

This theme also includes memories associated with school days, emphasizing the importance of academic socialization in shaping childhood experiences. For instance, we read memories of getting rewarded for behaving well in class, getting in trouble at school for squeezing orange juice, and crying on the first day of kindergarten because a friend didn't wave hello. School provides an early context for learning to navigate one's emotions in social settings, serving as an incubator for cultivating key social and emotional skills.[124]

Learning experiences are not limited to the schoolyard. There were several earliest memories of children testing boundaries around mischief, rules, and discipline. We read stories of breaking the "no running in the house" rule, the indignity of being put to bed against one's will, and the anxious anticipation of getting in trouble for eating a birthday candle or stealing a turtle. These memories shed light on the way children develop their identity in relation to authority figures.

From a young age, people display a wide range of reactions to authority, from the fear of non-compliance to the thrill of mischief and the spark of defiance. While it may test parents' patience at times, pushing limits with authority figures is an important part of young children's socio-emotional growth. It allows kids to explore their autonomy, develop moral reasoning, and understand their role in the world.[125-127] We all must find balance in life between asserting our independence to meet our personal needs and complying to maintain our relationships and community standing. The quest for that balance begins at a young age and is refined through a gradual process of trial and error, like two sides of a seesaw slowly finding their equilibrium.

Adaptation and Transformation

The theme of Adaptation and Transformation includes memories of surprises, accidents, transitions, and milestones. For example, we read memories of celebrating a fourth birthday, being surprised with a new bike on Christmas, preparing for a new sibling to be born, moving from a small village to a large town, and breaking one's front teeth on a coffee table. These memories highlight children's resilience and ability to navigate challenges, embrace new experiences, and find stability in the face of change.

Childhood is a period marked by the ebb and flow of inevitable shifts and transitions. Whether starting school for the first time or moving to a new neighborhood, children face common changes and coming-of-age milestones. Major life events that signify a total shift in context can make it easier to pinpoint our memories in time. For example, if you moved to a new house at age six, you know that all the memories of your first house were from that earlier time period.

Whether it is a significant life event, such as a change in family structure, or an everyday challenge, such as a canceled gymnastics class, each circumstance presents a unique opportunity for children to practice essential life skills like social adaptability, cognitive flexibility, problem- solving, and strength in the face of uncertainty.[122,128-130] When faced with unforeseen challenges, children demonstrate an innate ability to adapt and even thrive amidst change. Navigating unexpected events equips them with valuable skills needed to cope with life's inherent uncertainties. In essence, childhood becomes a training ground for resilience, laying the foundation for lifelong learning and growth.

What Makes Our First Memories So Special?

During the early to mid-20th century, psychoanalysts regarded a person's earliest memory as a window to the landscape of their emotional life.[131,132] Much like the opening scene of a movie sets the stage for what the film is about, the opening scene of our conscious life was viewed as a foundational marker of our character. As Rudolf Dreikurs phrased it (Dreikurs, 1935, p. 114[131]):

> The earliest memories of his childhood and his dreams can take him a long way on his voyage of discovery. They show the direction in which he tends to go. The earliest memories of childhood are always significant. They record experiences in response to which the child developed his characteristic attitude and there can be no doubt that each individual tries to justify his attitude by looking back to those experiences.

This charming viewpoint has largely fallen out of favor with modern scientists who assert that one's earliest memory is not remarkably different from other childhood memories.[133] Moreover, contemporary views hold that our first memories may not accurately represent actual events. Scientists believe it is rare to remember anything before age three.[22,36] However, in one large-scale study, nearly 40% of participants reported a first memory before the age of two. This was so scientifically implausible that the researchers labeled these memories as "fictional" or, at the very least, misdated.[134] This revelation sparked a series of disheartening news articles proclaiming that our earliest recollections are likely false.[135-137]

So, what can you conclude about *your* first memory? What

does it say about you? Does it mean anything at all? As a scientist, I'm interested in the theories and evidence that can help us better understand how memory develops and functions. But as a therapist and a romantic, I care much more about the personal meaning we draw from our own experiences. As novelist Vladimir Nabokov once said, "There can be no science without fancy, no art without facts." (Hannibal, 2013[138])

For most of us, our earliest memories *feel* significant and deeply meaningful.[139] So, does it really matter whether your first memory is "real" or simply a mosaic of reconstructed images and recycled family narratives? From my perspective, it is the connection to oneself and the creation of our own self-narrative that matter most.

I polled the respondents who shared their stories for the *Earliest Memories* collection to ask why they thought this particular memory stayed with them all those years. Some could not say why, but most people found the memory meaningful in some way. The most common reasons were because the emotion or sensation was so intense, the connection with the other person so profound, or the experience so novel. For some, it was simply the first time they really became aware of themselves or their surroundings.

Perhaps the most obvious reason why our earliest memories feel so special is because these memories are often deeply personal. No one has a memory *exactly* like yours. Maybe you are not the only person to recall riding the bus on your first day of school, but no one experienced it just like you. No one saw the scene through your eyes or traced the tiny hole in the vinyl seat with their little fingers just like you did. Our memories are unique to us, almost like a mental fingerprint, the first trace our human experience leaves on our consciousness.

Another reason our first memory can feel meaningful is due to familiarity. It is, by definition, the oldest thing we can consciously connect with. Of course, some people had never thought about their first memory until I asked them, but many have reflected on it over the years and turned the memory over in their mind again and again until it felt near and dear to them.

It is also possible that our earliest memory feels meaningful simply because it stands out to us. When we search our minds for our oldest memory, one moment may ring clearer than the others, leading us to conclude that it must be important. Our attention is naturally drawn to experiences that embody who we are. If multiple memories arise from the same period, we may connect most strongly with the one that resonates with our self-concept.

An earliest memory is a very intimate thing. Aside from perhaps a few close family members, it's unlikely that anyone in your current day-to-day life was there to witness it. Your earliest memory remains tucked away in a remote corner of your mind, something sacred that no one else can know unless you choose to share it. I am often amazed and honored that so many acquaintances and strangers shared something as precious as their first memory. At the same time, I am not wholly surprised. By writing a memory down, we transform it from a faint abstraction into something tangible. By speaking our first memory aloud, we multiply it—the memory now exists in the listener's mind as well. In this way, by sharing our memories, we find a way to immortalize ourselves.

Early memories are reminders of the formative years for which we hold a special fondness. In the next section, we explore the nature of nostalgia and what it means to long for the past.

The Nature of Nostalgia

Nostalgia is the desire to go back to an ideal time in one's life or a bygone era.[140,141] The word stems from the Greek *nóstos* (to return home) and *álgos* (pain), representing a kind of homesick longing for the past. Some view nostalgia as an unhealthy form of escapism. In the 17th century, nostalgia was even considered a disease that doctors attempted to treat with leeches and opium.[140,142] Indeed, being overly preoccupied with the past at the expense of engaging in the present moment can hinder our ability to function, learn, and grow.[142,143] However, there are adaptive ways to reminisce on the past without trying to reinstate it. Modern scientists have highlighted many positive influences of nostalgia, such as fostering a stronger sense of identity, reflecting on how we have changed over time, and maintaining social bonds through shared memories.[112]

Our archive of memories is both completely personal and universal at the same time. The stories we tell are part of a broader narrative that unites us as humans. Regardless of what you have experienced, we have all felt the upliftment of joy, hope, and awe, and the plunge of pain, confusion, and defeat. Through this collective emotional resonance, we discover comfort, empathy, and a sense of belonging in a world that can often feel challenging and isolating.

Nostalgia is a bittersweet feeling, a warm appreciation for bygone days tinged with the wistful acknowledgement that those golden times have slipped away. In Portugal, they have a beautiful concept called *sodade*, which captures the feeling of a deep, melancholic yearning for someone or something that has been lost. The Portuguese believe that a life fully lived has been marked with heartbreak, as joy is forever accompanied by an undercurrent of sorrow. So too, with memory. We keep the past with us, ever-present in our minds, hovering, ready to spring forth at the first glimpse of a baseball field or the slightest whiff of aftershave. We cannot summon the pleasant without also inviting the painful—it is our nature to live in two worlds. But it is in the meaning we make of our memories that the true beauty of life unfolds.

Closing Thoughts

Thank you for joining me on this journey through the labyrinth of memory. Together, we explored how memory develops and how it shapes our identity. We learned how memories persist, morph, and fade over time. We traveled through the landscape of emotional imprints and sensory impressions, brought colorfully to life in the collection of earliest childhood memories from across the world. Finally, we discovered common childhood themes that transcend cultural and generational boundaries, capturing the common experiences and milestones that encompass our collective existence.

Whether you choose to examine memory through the lens of science or nostalgia, I leave you with two resounding truths: 1) Our memories are not artifacts locked in a time capsule—they are ever-evolving stories that shape our identity and grow with us as we mature. 2) Memory is not a solitary endeavor, but rather a collective account of shared emotions as we travel through this human experience.

Memory, complex and imperfect, guides us through the

corridors of time, and in the meaning we make, our unique narrative unfolds—a story we continue to write with each passing moment. May this book be an invitation for you to revisit both the pleasure and pain of your childhood with a deep appreciation for the experiences that have made you uniquely "you."

"We now know that memories are not fixed or frozen, like Proust's jars of preserves in a larder, but are transformed, disassembled, reassembled, and recategorized with every act of recollection."

-Oliver Sacks, *Hallucinations*, 2012. p. 154[144]

Join the *Earliest Memories* Community

Share Your Earliest Memory

To read more memories from the collection and to share your earliest memory, visit the author's website at http://www.earliestmemoriesbook.com or email us at contact@earliestmemoriesbook.com.

Connect with Us on Social Media

- TikTok: @earliestmemoriesbook
- Instagram: @earliestmemoriesbook
- Facebook: @earliestmemoriesbook

Get the Word Out

Please help us get the word out so we can continue to grow the *Earliest Memories* community. Asking someone, **"What is your earliest memory?"** is a wonderful way to get to know someone better. Don't forget to share the *Earliest Memories* book website with your family and friends so they can add their memory to our collection too!

References

1. Nabokov, V. (1966). *Speak, memory: An autobiography revisited (Rev.)*. Putnam.

2. Gilboa, A. (2004). Autobiographical and episodic memory—one and the same? Evidence from prefrontal activation in neuroimaging studies. *Neuropsychologia, 42*(10), 1336–1349.

3. Lampinen, J., & Beike, D. (2015). *Memory 101* (The Psych 101 Series). Springer Publishing Company. https://doi.org/10.1891/9780826109255

4. Atkinson, R. C., & Shiffrin, R. M. (1968). Human memory: A proposed system and its control processes. In K. W. Spence, J. T. Spence (Eds.) *The psychology of learning and motivation. Volume 2*. Academic Press.

5. Tulving, E., & Thomson, D. M. (1973). Encoding specificity and retrieval processes in episodic memory. *Psychological Review, 80*, 352–373.

6. Tulving, E. (1974). Cue-dependent forgetting. *American Scientist, 62*, 74–82.

7. Metcalfe, J. (2005). CHARM^2: A multimodular model of human memory. In A. Parker, E. L. Wilding, T. J. Bussey (Eds.), *The cognitive neuroscience of memory: Encoding and retrieval.* (pp. 283-305). Psychology Press, Taylor and Francis Group. https://doi.org/10.4324/9780203989388

8. Eich, J. M. (1982). A composite holographic associative

recall model. *Psychological Review, 89*(6), 627–661. https://doi.org/10.1037/0033295X.89.6.627

9. Eichenbaum, H. (2001). The hippocampus and declarative memory: Cognitive mechanisms and neural codes. *Behavioural Brain Research, 27*(1-2), 199–207. https://doi.org/10.1016/S0166-4328(01)00365-5

10. Eichenbaum, H. (2011). The cognitive neuroscience of memory: An introduction. *In Cellular mechanisms of memory: Complex circuits.* Essay, Oxford University Press. https://doi.org/10.1093/acprof:oso/9780199778614.003.0003

11. Jesse, J. L., & Richard, E. B. (2018). The synaptic theory of memory: A historical survey and reconciliation of recent opposition. *Frontiers in Systems Neuroscience, 12.* https://doi.org/10.3389/fnsys. 2018.00052

12. Korte, M., & Schmitz, D. (2016). Cellular and system biology of memory: Timing, molecules, and beyond. *Physiological Reviews, 96*(2), 647–693. https://doi.org/10.1152/physrev. 00010.2015

13. Neisser, U. (1967). *Cognitive psychology.* Appleton-Century-Crofts.

14. Schacter, D. L., & Addis, D. R. (2007). The cognitive neuroscience of constructive memory: remembering the past and imagining the future. *Philosophical Transactions of the Royal Society B: Biological Sciences, 362*(1481), 773–786. https://doi.org/10.1098/rstb. 2007.2087

15. Godden, D. R., & Baddeley, A. D. (1975) Context-dependent memory in two natural environments: On

land and underwater. *British Journal of Psychology, 66*, 325-331. https://doi. org/10.1111/j.2044-8295.1975. tb01468.x

16. Folkerts, S., Rutishauser, U., & Howard, M. W. (2018). Human episodic memory retrieval is accompanied by a neural contiguity effect. *The Journal of Neuroscience: The Official Journal of the Society for Neuroscience, 38*(17), 4200–4211. https://doi.org/10. 1523/JNEUROSCI.231217. 2018

17. Suddendorf, T., & Corballis, M. C. (1997). Mental time travel and the evolution of the human mind. *Genetic, Social, and General Psychology Monographs, 123*(2), 133–167.

18. Tulving, E. (1985). Memory and consciousness. *Canadian Psychology, 26*, 1–12. https://doi.org/10.1037/ h0080017

19. Bartlett F. (1932). *Remembering.* Cambridge University Press.

20. Radvansky, G. A., & Tamplin, A. K. (2012). Memory. In V. S. Ramachandran (Ed.), *Encyclopedia of human behavior* (2nd ed., pp. 585–592). Elsevier. https://doi.org/ 10.1016/ B978-0-12-375000-6.00229-9

21. Slotnick, S. D. (2017). *Cognitive neuroscience of memory.* Cambridge fundamentals of neuroscience in psychology. Cambridge University Press.

22. Baddeley, A. D. (1999). *Essentials of human memory.* Psychology Press.

23. Walker, M. P., & Stickgold, R. (2004). Sleep-dependent

learning and memory consolidation. *Neuron, 44*(1), 121–33.

24. Anderson, M. C., & Hulbert, J. C. (2021). Active forgetting: Adaptation of memory by prefrontal control. *Annual Review of Psychology, 72*, 1–36. https://doi.org/10.1146/annurev-psych-072720-094140

25. Loftus, E. F., & Palmer, J. C. (1974). Reconstruction of automobile destruction: An example of the interaction between language and memory. *Journal of Verbal Learning & Verbal Behavior, 13*(5), 585–589. https://doi.org/10.1016/S0022-5371(74)80011-3

26. Patihis, L., Frenda, S., LePort, A., Petersen, N., Nichols, R., Stark, C., . . . Loftus, E. (2013). False memories in highly superior autobiographical memory individuals. *Proceedings of the National Academy of Sciences of the United States of America, 110*(52), 20947-20952. https://doi.org/10.1073/pnas.1314373110

27. Smeets, T., Telgen, S., Ost, J., Jelicic, M., & Merckelbach, H. (2009). What's behind crashing memories? Plausibility, belief and memory in reports of having seen non-existent images. *Applied Cognitive Psychology, 23,* 1333–1341. https://doi.org/10.1002/acp.1544

28. Ost, J., Vrij, A., Costall, A., & Bull, R. (2002). Crashing memories and reality monitoring: Distinguishing between perceptions, imaginations and 'false memories'. *Applied Cognitive Psychology, 16,* 125–134. https://doi.org/10.1002/acp.779

29. LePort, A. K. R., MaHeld, A. T., Dickinson-Anson, H.,

Fallon, J. H., Stark, C. E. L., Kruggel, F., Cahill, L., & McGaugh, J. L. (2012). Behavioral and neuroanatomical investigation of highly superior autobiographical memory (HSAM). *Neurobiology of Learning and Memory, 98*(1), 78–92. https://doi.org/ 10.1016/j.nlm.2012.05.002

30. Finkelstein, S. (2011, June 24). *Understanding the gift of endless memory* [Radio broadcast]. CBS Interactive, *60 Minutes*. https://www.cbsnews.com/news/understanding-the-gift-of-endless-memory/

31. Spiegel, A. (2013, December 27). *When memories never fade, the past can poison the present.* [Radio broadcast]. NPR, *All Things Considered.* https://www.npr.org/sections/healthshots/2013/12/18/255285479/when-memories-never-fade-the-past-can-poison-the-present

32. Lacy, J. W., & Stark, C. E. L. (2013). The neuroscience of memory: Implications for the courtroom. *Nature Reviews Neuroscience, 14*(9), 649–658. https://doi.org/10.1038/nrn3563

33. Taylor, S. E. (1989). Positive illusions: Creative self-deception and the healthy mind. Basic Books.

34. Heine, S. J., Lehman, D. R., Markus, H. R., & Kitayama, S. (1999). Is there a universal need for positive self-regard? *Psychological Review, 106*(4), 766–794. https://doi.org/10.1037/0033-295X.106.4.766

35. Wilson, A., & Ross, M. (2003). The identity function of autobiographical memory: Time is on our side. *Memory, 11*(2), 137–149. https://doi.org/10.1080/741938210

36. Rubin, D. (2000). The distribution of early childhood memories. *Memory, 8*(4), 265–269. https://doi.org/10.1080/096582100406810

37. Bruce, D., Dolan, A., & Phillips-Grant, K. (2000). On the transition from childhood amnesia to the recall of personal memories. *Psychological Science, 11,* 360–364. https://doi.org/10.1111/1467-9280.0027

38. Bauer, P. J., & Larkina, M. (2014). Childhood amnesia in the making: Different distributions of autobiographical memories in children and adults. *Journal of Experimental Psychology: General, 143*(2), 597–611. https://doi.org/10.1037/a0033307

39. Pathman, T. & Bauer, P.J. (2020). Memory and Early Brain Development. In R. E. Tremblay, M. Boivin, R. DeV. Peters, (Eds.), *Encyclopedia on Early Childhood Development.* Centre of Excellence for Early Childhood Development. http://www.child-encyclopedia.com/brain/according-experts/memory-and-early-brain-development.

40. Piaget, J. (1954). *The construction of realty in the child.* (M. Cook, Trans.). Basic Books.

41. Vygotsky, L. (1986/1934). *Thought and language.* MIT Press.

42. Nelson, K., & Fivush, R. (2020). The development of autobiographical memory, autobiographical narratives, and autobiographical consciousness. *Psychological Reports, 123*(1), 71–96. https://doi.org/10.1177/0033294119852574

43. Geng, F., Canada, K., & Riggins, T. (2018). Age- and performance-related differences in encoding during early childhood: Insights from event-related potentials. *Memory, 26*(4), 451–461. https://doi.org/10.1080/09658211.2017.1366526

44. Howe, M. L., & O'Sullivan, J. T. (1997). What children's memories tell us about recalling our childhoods: A review of storage and retrieval processes in the development of long-term retention. *Developmental Review, 17*(2), 148–204.

45. Bruner, J. S. (1978). The role of dialogue in language acquisition. In A. Sinclair, R., J. Jarvelle, and W. J.M. Levelt (Eds.) *The child's concept of language.* Springer-Verlag.

46. Fivush, R. (2019). Sociocultural developmental approaches to autobiographical memory. *Applied Cognitive Psychology, 33*(4), 489–497. https://doi.org/10.1002/acp.3512

47. Hedrick, A. M., Haden, C. A., & Ornstein, P. A. (2009). Elaborative talk during and after an event: conversational style influences children's memory reports. *Journal of Cognition and Development, 10*(3), 188–209.

48. Sales, J. M. D., Fivush, R., & Peterson, C. (2003). Parental reminiscing about positive and negative events. *Journal of Cognition and Development, 4*(2), 185–209.

49. Salmon, K., & Reese, E. (2016). The benefits of reminiscing with young children. *Current Directions in Psychological Science, 25*(4), 233–238. https://doi.org/10.

1177/0963721416655100

50. Locke, J. (1948/1690). An essay concerning human understanding. In W. Dennis (Ed.), *Readings in the history of psychology* (pp. 55–68). Appleton-Century-Crofts. https://doi.org/10.1037/11304-008

51. Çili, S., & Stopa, L. (2019). *Autobiographical memory and the self: Relationship and implications for cognitive-behavioural therapy.* Milton: Routledge.

52. Halbwachs, M. (1950). *La mémoire collective* [Collective memory]. (Ser. Bibliothèque de sociologie contemporaine). Presses Universitaires de France.

53. National Geographic Society. Storytelling and cultural traditions. Article. Retrieved Online 6/7/23 from: https://education.nationalgeographic.org/resource/storytelling-and-cultural-traditions/

54. Assmann, J. (1992). *Das kulturelle gedächtnis, schrik, erinnerung und politische identität in frühen hochkulturen* [Cultural memory: Writing, memory, and political identity in early civilizations]. C.H. Beck.

55. Assmann, J. (2011). *Cultural memory and western civilization: Functions, media, archives.* Cambridge University Press.

56. National Geographic Society. Cultural memory. Encyclopedic Entry. Retrieved Online 3/4/23 from: https://education.nationalgeographic.org/resource/cultural-memory/

57. Frey, W. H., & Langseth, M. (1985). *Crying: The mystery of tears.* Winston Press.

58. Gračanin, A., Bylsma, L.M., Vingerhoets, A. J. J. M. (2014). Is crying a self-soothing behavior? *Frontiers in Psychology, 5,* 502. https://doi.org/10.3389/fpsyg.2014.00502

59. Orloff, J. (2010). The health benefits of tears. *Psychology Today.* https://www.psychologytoday.com/intl/blog/emotional-freedom/201007/the-health-benefits-tears

60. Miceli, M., & Castelfranchi, C. (2018). Reconsidering the differences between shame and guilt. *Europe's Journal of Psychology, 14*(3), 710–733. https://doi.org/10.5964/ejop.v14i3.1564

61. Dolcos, F., Katsumi, Y., Weymar, M., Moore, M., Tsukiura, T., & Dolcos, S. (2017). Emerging Directions in Emotional Episodic Memory. *Frontiers in Psychology, 8,* 286040. https://doi.org/10.3389/fpsyg.2017.01867

62. Dolcos, F., LaBar, K. S., & Cabeza, R. (2004). Interaction between the amygdala and the medial temporal lobe memory system predicts better memory for emotional events. *Neuron, 42*(5), 855–863. https://doi.org/10.1016/s0896-6273(04)00289-2

63. Hu, H., Real, E., Takamiya, K., Kang, M.-G., Ledoux, J., Huganir, R. L., & Malinow, R. (2007). Emotion enhances learning via norepinephrine regulation of ampa-receptor trafficking. *Cell, 131*(1), 160–173. https://doi.org/10.1016/j.cell.2007.09.017

64. McGaugh, J.L., McIntyre, C.K., and Power, A.E. (2002). Amygdala modulation of memory consolidation: Interaction with other brain systems. *Neurobiology of Learning and Memory, 78,* 539–552.

https://doi.org/10.1006/nlme.2002.4082

65. Aly, M., & Turk-Browne, N. B. (2016). Attention promotes episodic encoding by stabilizing hippocampal representations. *Proceedings of the National Academy of Sciences of the United States of America, 113*(4), 420–429. https://doi.org/10.1073/pnas.1518931113

66. Talarico, J. M., LaBar, K. S., & Rubin, D. C. (2004). Emotional intensity predicts autobiographical memory experience. *Memory & Cognition, 32*(7), 1118–1132. https://doi.org/10.3758/BF03196886

67. Dolcos, F., LaBar, K. S., & Cabeza, R. (2005). Remembering one year later: Role of the amygdala and the medial temporal lobe memory system in retrieving emotional memories. *Proceedings of the National Academy of Sciences, 102*(7), 2626–2631. https://doi.org/10.1073/pnas.0409848102

68. Sheldon, S., & Donahue, J. (2017). More than a feeling: Emotional cues impact the access and experience of autobiographical memories. *Memory & Cognition, 45*(5), 731–744. https://doi.org/10.3758/s13421-017-0691-6

69. Riegel, M., Wierzba, M., Grabowska, A., Jednoróg Katarzyna, & Marchewka, A. (2016). Effect of emotion on memory for words and their context. *Journal of Comparative Neurology, 524*(8), 1636–1645. https://doi.org/10.1002/cne.23928

70. Eich, E., Macaulay, D., & Ryan, L. (1994). Mood dependent memory for events of the personal past. *Journal of Experimental Psychology, 123*(2), 201–201. https://doi.org/10.1037/0096-3445.123.2.201

71. Levine, L. J., & Pizarro, D. A. (2004). Emotion and memory research: A grumpy overview. *Social Cognition, 22*(5), 530–554. https://doi.org/10.1521/soco.22.5.530.50767

72. Levine, L. J., & Bluck, S. (2004). Painting with broad strokes: Happiness and the malleability of event memory. *Cognition and Emotion, 18*(4), 559–574. https://doi. org/10.1080/02699930341000446

73. Speer, M. E., Bhanji, J. P., & Delgado, M. R. (2014). Savoring the past: positive memories evoke value representations in the striatum. *Neuron, 84*(4), 847–856. https://doi.org/10.1016/j.neuron.2014.09.028

74. Walker, W. R., Skowronski, J. J., & Thompson, C. P. (2003). Life is pleasant—and memory helps to keep it that way! *Review of General Psychology, 7*(2), 203–210. https://doi.org/10.1037/1089-2680.7.2.203

75. Siedlecki, K.L., Hicks, S., Kornhauser, Z.G. (2015). Examining the positivity effect in autobiographical memory across adulthood. *The International Journal of Aging and Human Development, 80*(3), 213–232. https://doi.org/10.1177/0091415015590311

76. Reed, A. E., Chan, L., & Mikels, J. A. (2014). Meta-analysis of the age-related positivity effect: Age differences in preferences for positive over negative information. *Psychology and Aging, 29*(1), 1–15. https://doi.org/10.1037/a0035194

77. Kensinger, E. A. (2009). Remembering the details: effects of emotion. *Emotion Review, 1*(2), 99–113. https://doi. org/10.1177/1754073908100432

78. Payne, J. D., Jackson, E. D., Hoscheidt, S., Ryan, L., Jacobs, W. J., & Nadel, L. (2007). Stress administered prior to encoding impairs neutral but enhances emotional long-term episodic memories. *Learning & Memory, 14*(12), 861–868. https://doi.org/10.1101/lm.743507

79. Steig, P. (Host). (2019, September 13). Our Emotional Memory, with Dr. Elizabeth Phelps [Audio podcast episode]. In *This Is Your Brain*. This Is Your Brain. https://thisisyourbrain.com/2022/01/s3-episode-2-our-emotional-memory/

80. Levine, P. A., & Frederick, A. (1997). *Waking the tiger: Healing trauma: The innate capacity to transform overwhelming experiences*. North Atlantic Books.

81. Payne, P., Levine, P. A., & Crane-Godreau, M. A. (2015). Somatic experiencing: using interoception and proprioception as core elements of trauma therapy. *Frontiers in Psychology, 6*. https://doi.org/10.3389/fpsyg.2015.00093

82. Van der Kolk, B. A. (2015). *The body keeps the score: Brain, mind, and body in the healing of trauma*. Penguin Books.

83. Mørkved, N., Hartmann, K., Aarsheim, L. M., Holen, D., Milde, A. M., Bomyea, J., & Thorp, S. R. (2014). A comparison of narrative exposure therapy and prolonged exposure therapy for PTSD. *Clinical Psychology Review, 34*(6), 453–467. https://doi.org/10.1016/j.cpr.2014.06.005

84. Van der Kolk, B. A., & Fisler, R. (1995). Dissociation and the fragmentary nature of traumatic memories:

Overview and exploratory study. *Journal of Traumatic Stress, 8*(4), 505–525. https://doi.org/10.1007/BF02102887

85. Eacott, M. J. (1999). Memory for the events of early childhood. *Current Directions in Psychological Science, 8*(2), 46-48.

86. Rochat, P. (2003). Five levels of self-awareness as they unfold early in life. *Consciousness and Cognition: An International Journal, 12*(4), 717–731. https://doi.org/10.1016/S1053-8100(03)00081-3

87. Andrews, G., Murphy, K., Dunbar, M. (2020). Self-referent encoding facilitates memory binding in young children: New insights into the self-reference effect in memory development. *Journal of Experimental Child Psychology, 198,* 104919. https://doi.org/10.1016/j.jecp.2020.104919

88. Symons, C. S., & Johnson, B. T. (1997). The self-reference effect in memory: A meta-analysis. *Psychological Bulletin, 121*(3), 371– 394. https://doi.org/10.1037/0033-2909.121.3.371

89. Huang, N., & Elhilali, M. (2020). Push-pull competition between bottom-up and top-down auditory attention to natural soundscapes. *ELife, 9.* https://doi.org/10.7554/eLife.52984

90. Erickson, L. C., & Newman, R. S. (2017). Influences of background noise on infants and children. *Current Directions in Psychological Science, 26*(5), 451–457. https://doi. org/10.1177/0963721417709087

91. Werner, L. A. (2007). Issues in human auditory development. *Journal of Communication Disorders, 40*(4), 275–283.

92. Blasi, A., Mercure, E., Lloyd-Fox, S., Thomson, A., Brammer, M., Sauter, D., Deeley, Q., Barker, G. J., Renvall, V., Deoni, S., Gasston, D., Williams, S. C. R., Johnson, M. H., Simmons, A., & Murphy, D. G. M. (2011). Early specialization for voice and emotion processing in the infant brain. *Current Biology, 21*(14), 1220–1224. https://doi.org/10.1016/j.cub.2011.06.009

93. Amorim, M., Anikin, A., Mendes, A. J., Lima, C. F., Kotz, S. A., & Pinheiro, A. P. (2021). Changes in vocal emotion recognition across the life span. *Emotion, 21*(2), 315–325. https://doi.org/10.1037/emo0000692

94. Burrell, L. V., Johnson, M. S., & Melinder, A. (2016). Children as earwitnesses: Memory for emotional auditory events. *Applied Cognitive Psychology, 30*(3), 323–331. https://doi.org/10.1002/acp.3202

95. Proust, M. (1960/1922). *In search of lost time: Swann's way.* (C. K. Scott Moncrieff, Trans.). Chatto & Windus.

96. Chu, S., and Downes, J. J. (2000). Long live Proust: The odour-cued autobiographical memory bump. *Cognition, 75*, B41–B50. https://doi.org/10.1016/s0010-0277(00)00065-2

97. Chu, S., & Downes, J. J. (2002). Proust nose best: Odors are better cues of autobiographical memory. *Memory & Cognition, 30*(4), 511–518. https://doi.org/10.3758/BF03194952

98. Herz, R. (2007). *The scent of desire: Discovering our enigmatic sense of smell.* HarperCollins Publishers.

99. Castillo, M. (2014). The complicated equation of smell, flavor, and taste. *AJNR. American Journal of Neuroradiology, 35*(7), 1243–5. https://doi.org/10.3174/ajnr.A3739

100. Ernst, A., Bertrand, J. M. F., Voltzenlogel, V., Souchay, C., & Moulin, C. (2021). The Proust machine: What a public science event tells us about autobiographical memory and the five senses. *Frontiers in Psychology, 11,* 623910. https://doi.org/10.3389/fpsyg.2020.623910

101. Allen, J. S. (2012). *The omnivorous mind: Our evolving relationship with food.* Harvard University Press. https://doi.org/10.4159/harvard.9780674064737

102. Lisman, J. E., & Grace, A. A. (2005). The hippocampal-VTA loop: Controlling the entry of information into long-term memory. *Neuron, 46*(5), 703–713. https://doi.org/10.1016/j.neuron.2005.05.002

103. Shohamy, D., & Adcock, R. A. (2010). Dopamine and adaptive memory. *Trends in Cognitive Sciences, 14*(10), 464–472. https://doi.org/10.1016/j.tics.2010.08.002

104. Gadelha, M. J. N., da Silva, J. A., de Andrade, M. J. O., Viana, D. N. D. M., Calvo, B. F., & dos Santos, N. A. (2013). Haptic memory and forgetting: A systematic review. *Estudos De Psicologia (Natal), 18*(1), 131–136. https://doi.org/10.1590/S1413-294X2013000100021

105. Lederman, S. J., & Klatzky, R. L. (2009). *Attention, Perception, & Psychophysics, 71*(7), 1439–1459.

https://doi.org/10.3758/app.71.7.1439

106. Sheets-Johnstone, M. (2007). Kinesthetic memory. *Theoria Et Historia Scientiarum, 7*(1), 69–69. https://doi.org/10.12775/ths.2003.005

107. Zhang, M., Zhang, Y., & Kong, Y. (2019). Interaction between social pain and physical pain. *Brain Science Advances, 5*(4), 265– 273. https://doi.org/10.26599/BSA.2019.9050023

108. DeWall, N. C., MacDonald, G., Webster, G., Masten, C., Baumeister, R., Powell, C., Combs, D., Schurtz, D., Stillman, T., Tice, D., & Eisenberger, N. (2010). Acetaminophen reduces social pain: Behavioral and neural evidence. *Psychological Science, 21*(7), 931–937.

109. Adamczyk, W. M., Farley, D., Wiercioch-Kuzianik, K., Bajcar, E. A., Buglewicz, E., Nastaj, J., Gruszka, A., & Bąbel, P. (2019). Memory of pain in adults: A protocol for systematic review and meta- analysis. *Systematic Reviews, 8.* https://doi.org/10.1186/s13643-019-1115-4

110. Von Baeyer, C. L., Marche, T. A., Rocha, E. M., & Salmon, K. (2004). Children's memory for pain: Overview and implications for practice. *The Journal of Pain, 5*(5), 241–249.

111. Pavlova, M., Kennedy, M., Lund, T., Jordan, A., & Noel, M. (2022). Let's (not) talk about pain: Mothers' and fathers' beliefs regarding reminiscing about past pain. *Frontiers in Pain Research, 3,* 890897. https://doi.org/10.3389/fpain.2022.890897

112. Batcho, K. I., Nave, A. M., & DaRin, M. L. (2011). A

retrospective survey of childhood experiences. *Journal of Happiness Studies: An Interdisciplinary Forum on Subjective Well-Being, 12*(4), 531– 545. https://doi.org/10.1007/s10902-010-9213-y

113. Ainsworth, M. D. S. (1967). *Infancy in Uganda: Infant care and the growth of love*. Baltimore: Johns Hopkins University Press.

114. Bowlby, J. (1969). *Attachment and loss: Vol. 1. Attachment.* Basic Books.

115. Chopik, W. J., & Edelstein, R. S. (2019). Retrospective memories of parental care and health from mid- to late life. *Health Psychology: Official Journal of the Division of Health Psychology, American Psychological Association, 38*(1), 84–93. https://doi.org/10.1037/hea0000694

116. Pinquart, M. (2022). Associations of self-esteem with attachment to parents: A meta-analysis. *Psychological Reports, 126*(5), 2101– 2118. https://doi.org/10.1177/00332941221079732

117. Schore, J. R., & Schore, A. N. (2008). Modern attachment theory: the central role of affect regulation in development and treatment. *Clinical Social Work Journal, 36*(1), 9–20. https://doi.org/10.1007/s10615-007-0111-7

118. Fonagy, P., Gergely, G., Jurist, E. L., & Target, M. (2002). *Affect Regulation, Mentalization and the Development of the Self.* New York: Other Press.

119. Lewicka, M. (2011). Place attachment: How far have we come in the last 40 years? *Journal of Environmental*

Psychology, 31(3), 207–230. https://doi.org/10.1016/j.jenvp.2010.10.001

120. Piaget, J. (1962). *Play, dreams, and imitation in childhood.* New York: Norton.

121. Rousseau, J.-J. (1762). *Émile, or on education.*

122. Vygotsky, L. S. (1978). *Mind in society: The development of higher psychological processes.* Cambridge, MA: Harvard University Press.

123. Petrović-Sočo, B. (2013). Symbolic play of children at an early age. *Croatian Journal of Education, 16,* 235–251. https://doi.org/10.15516/cje.v16i0.1045

124. Jones, S. M., Barnes, S. P., Bailey, R., & Doolittle, E. J. (2017). Promoting social and emotional competencies in elementary school. *The Future of Children, 27*(1), 49–72.

125. Erikson, E. (1950). *Childhood and society.* W.W. Norton & Company.

126. Graves, S., & Larkin, E. (2006). Lessons from Erikson: A look at autonomy across the lifespan. *Journal of Intergenerational Relationships, 4*(2), 61–71.

127. Kohlberg, L. (1981). *Essays on moral development* (1st ed.). Harper & Row.

128. Antony, E. M. (2022). Framing childhood resilience through Bronfenbrenner's ecological systems theory: A discussion paper. *Cambridge Educational Research e-Journal, 9,* 244-257. https://doi.org/10.17863/CAM.90564

129. Bonanno, G. A. (2005). Resilience in the face of potential trauma. *Current Directions in Psychological Science,*

14(3), 135–138.

130. Dewey, J. (1938). *Experience and education.* New York: Macmillan.

131. Dreikurs, R. (1935). *An introduction to individual psychology.* Routledge.

132. Saul, L. J., Snyder, T. R., & Sheppard, E. (1956). On earliest memories. *The Psychoanalytic Quarterly, 25*(2), 228–237. https://doi.org/10.1080/21674086.1956.11926024

133. Kangaslampi, S. (2023). Earliest versus other autobiographical memories of school-age children. *Current Psychology.* https://doi.org/10.1007/s12144-023-04377-8

134. Akhtar, S., Justice, L. V., Morrison, C. M., & Conway, M. A. (2018). Fictional first memories. *Psychological Science, 29*(10), 1612–1619. https://doi.org/10.1177/0956797618778831

135. Chodosh, S. (2018, July). Your first memory probably isn't yours, no matter how real it seems. *Popular Science.* https://www.popsci.com/first-memory-false-neuroscience/

136. Griffiths, S. (2019, May). Can you trust your earliest childhood memories? *BBC.* https://www.bbc.com/future/article/20190516-why-you-cannot-trust-your-earliest-childhood-memories

137. Reinberg, S. (2018, July). Your earliest childhood memories may be false. *CBS News.* https://www.cbsnews.com/news/your-earliest-childhood-memories-

maybe-false/

138. Hannibal, M. E. (2013, December). Speak, Butterfly. *Nautilus*. https://nautil.us/speak-butterfly-234692/

139. Hutmacher, F., & Morgenroth, K. (2022). The beginning of the life story: The meaning of the earliest autobiographical memory from an adult perspective. *Applied Cognitive Psychology, 36*(3), 612–622. https://doi.org/10.1002/acp.3948

140. Boym, S. (2001). *The future of nostalgia*. Basic Books.

141. Random House. (1966). *Random House dictionary of the English Language*. Random House.

142. Peters, R. (1985). Reflections on the origin and aim of nostalgia. *Journal of Analytical Psychology, 30*, 135–148. https://doi.org/10.1111/j.1465-5922.1985.00135.x

143. Batcho, K. I. (2013). Nostalgia: The bittersweet history of a psychological concept. *History of Psychology, 16*(3), 165–176. https://doi.org/10.1037/a0032427

144. Sacks, O. (2012). *Hallucinations* (1st American ed.). Alfred A. Knopf.

About the Author

Dr. Charlotte Chun was born and spent her early childhood in the rural town of East Montpelier, Vermont. Her earliest memory from age four:

I am down around the side of my childhood home in Vermont. I am sitting on the ground. In my mind, it is dusty. I'm partaking in one of my favorite activities: eating gooseberries off the plant before they are ripe. I pluck a small, green berry and pop it in my mouth. I feel the sharpness on my tongue from the juice.

Dr. Chun grew up and became focused on other pursuits, leaving her less time to devote to eating gooseberries. She earned her Bachelor of Science in Psychology at Louisiana State University, Master of Science in Applied Cognition and Neuroscience at Université Paul Sabatier—Toulouse III, and Doctorate of Philosophy in Clinical Psychology at University of North Carolina at Greensboro, and is a licensed psychologist.

Her career has taken her on a dynamic path from studying the brain and human behavior to better understanding the health care market, empowering clients towards authentic expression through style and branding (www.drchuncoaching.com), writing in the personal development and wellness space, and providing psychotherapy to support psychedelic medicine, existential exploration, and somatic healing.